KU-054-254

Octopus

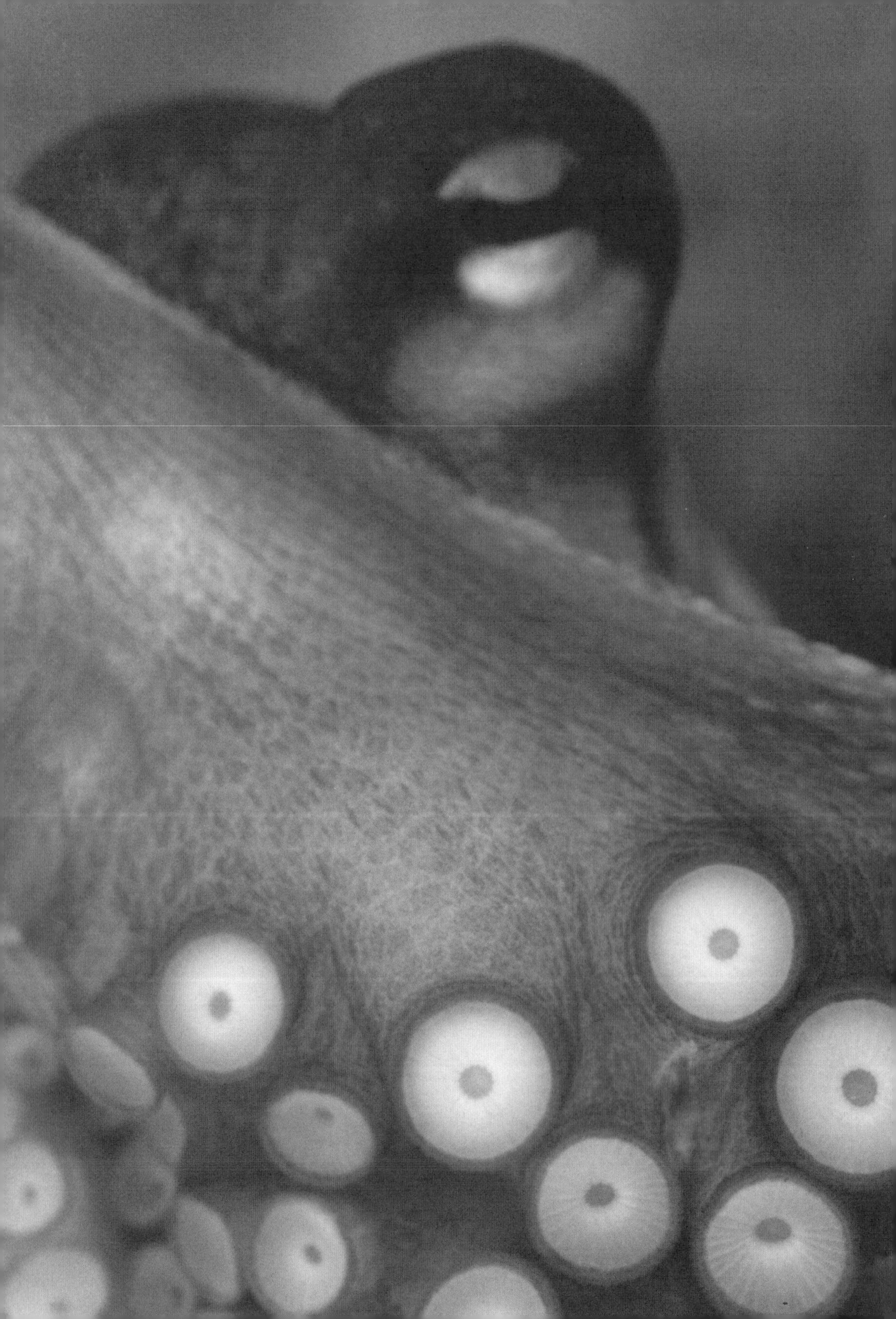

Octopus

The Ocean's Intelligent Invertebrate

Jennifer A. Mather,
Roland C. Anderson,
and James B. Wood

Timber Press
Portland • London

Published in 2010 by Timber Press, Inc.

The Haseltine Building
133 S.W. Second Avenue, Suite 450
Portland, Oregon 97204-3527
timberpress.com

2 The Quadrant
135 Salusbury Road
London NW6 6RJ
timberpress.co.uk

Third printing 2013
Printed in China

Library of Congress Cataloging-in-Publication Data

Mather, Jennifer A.
Octopus : the ocean's intelligent invertebrate / Jennifer A. Mather, Roland C. Anderson, and James B. Wood. — 1st ed.
p. cm.
Includes bibliographical references and index.
ISBN 978-1-60469-067-5
1. Octopuses. I. Anderson, Roland C. II. Wood, James B. III. Title.
QL430.3.O2M38 2010
594'.56—dc22

2009041311

A catalog record for this book is also available from the British Library.

For Lynn

Contents

Preface

Octopuses are intriguing and unpredictable, and our decades of studying them have given us truly exciting experiences. In fact, we three authors could have a competition for "peak octopus experience." Was it when Roland watched the giant Pacific octopus (*Enteroctopus dofleini*) named Olive (she mated with Popeye) tend her eggs until they hatched out in Puget Sound? Or was it when James, as a teenager, thought his self-caught octopus had "escaped" from the tank for three days, and then found it still in the tank but underneath the gravel filter plate? Then there was the event when Jennifer first witnessed tool use by the common octopuses (*Octopus vulgaris*) of Bermuda. Or maybe it was when Roland, still skeptical about octopus play, phoned Jennifer long-distance after watching an octopus blow jets of water at a floating pill bottle, causing the object to make circuits in its aquarium tank, and said, "She's bouncing the ball!" Perhaps it was when James managed to raise the elusive spoon-arm octopus (*Bathypolypus arcticus*) for the first time. For all of us, it was also when we looked directly at a camouflaged octopus, realized it looked exactly like the rock behind it, and wondered how the animal could do that. In this book, we share many of these captivating experiences with our readers.

Octopuses are found in most of the habitats in the ocean, and they are an important part of the sea's complex web of life. They do countless interesting things, and in the process challenge how we think about such issues as personality, intelligence, or play, thereby revealing much to us about what it means to be a living being on the earth. But most important, octopuses are wondrous to behold.

The three of us have studied the octopus and its behavior, in the lab and in the field at different locations on the planet, such as Banyuls on the southern coast of France, the islands of Bermuda, the small Caribbean island of Bonaire, and Hawaii. We all started out as what James calls "old-fashioned naturalists," walking and swimming along the shoreline and getting acquainted with what lives there—Jennifer on Vancouver Island, British Columbia, Roland on Puget Sound, Washington, and James in Florida. We started our university educations in biology, getting a foundation in the basics of marine animals, but our paths and major interests diverged somewhat after that.

In graduate school, Jennifer started studying the Caribbean pygmy octopus (*Octopus joubini/mercatoris*) but switched to psychology, adding an emphasis on theory to her marine biology background. As a university professor, she has explored theoretical ideas in areas such as octopus foraging strategies, personalities, intelligence, and consciousness, and she covers these topics in this book. Roland has worked for the Seattle Aquarium for over thirty years, getting a good grounding in issues of animal care. He's written papers about enrichment for octopuses in captivity, and, in a later part-time doctoral program, he focused on the sand-digging behaviors of stubby squid (*Rossia pacifica*). His background in these practical areas shows when he discusses many dimensions of octopus life in this book. James completed a doctoral program in biology and oceanography, accomplishing the technically demanding task of keeping the deep-sea spoonarm octopus and evaluating influences on lifespan. He's on the staff at the Aquarium of the Pacific, in Long Beach, California. Besides becoming an expert on keeping octopuses in aquariums and writing this book's section on the topic, he became interested in the Internet mode of communication, and formed the Web sites The Cephalopod Page and CephBase about living cephalopods. James is also an excellent underwater photographer, and many of the pictures in the book are his.

Together, we hope to help readers understand more about the mysterious marine world and the fascinating octopuses that live in it.

Acknowledgments

We would like to thank Roy Caldwell, John Forsythe, NOAA, Leo Shaw, Barry Shuman, Abel Valdiva, and Stuart Westmorland for providing photographs. We would also like to thank Elena Hannah, Sandra Palm, David Sinn, Kimberley Zeeh, and Brandi Walker for reading and commenting on book chapters. We also would like to thank Leanne Wehlage-Ellis for word processing. Foremost, we would like to acknowledge the octopuses for their inspiration.

Introduction
Meet the Octopus

Octopuses are amazing animals. They can change color, texture, and shape. They have three hearts pumping blue copper-based blood, and are jet powered. They can squeeze through the tiniest of cracks and disappear behind a cloud of ink. Octopuses adapt to new situations, solve problems, learn techniques, and are curious about their surroundings. They are considered by many to be the most intelligent of invertebrates. They have the manipulative ability to get into fishermen's crab traps, eat the crabs, and get out again. They can dismantle the aquariums they are kept in, and escape. They can even grow a new arm when one is bitten off.

The octopus is a mollusk, like clams and snails. Through evolution, it has lost the confining but protective shell. Since it is not hidden in the shell, the muscular mantle wrapped around the animals' insides is free to set up jet propulsion—it can contract to shoot water through its flexible funnel (see plate 1). With jet propulsion, the octopus can move through the water, shove aside sand and small rocks to clean out a home, and jet at annoyances like scavenging fish and pesky researchers. The octopus's head is usually raised up high so it can see well through its two lens-type eyes, which are very much like the eyes of mammals and birds except the pupil is a horizontal slit instead of round. The head is surrounded by eight arms that have flexible skin webs between them, and the mouth is underneath at the center of the arms. To name the octopus arms, we divide them into left and right arms by their position compared to a line down the middle of the animal. If an octopus is facing forward with its mantle to the rear, it has four arms on its right and four on the left. These are counted as R1, R2, L1, L2, and so on.

The fluid in the octopus's mantle cavity is an inside part of the outside ocean, vital for respiration and removal of waste. Water circulates around in it as in other mollusks, and its contraction is parallel to our breathing. The tubelike funnel directs the outgoing water after it has passed over

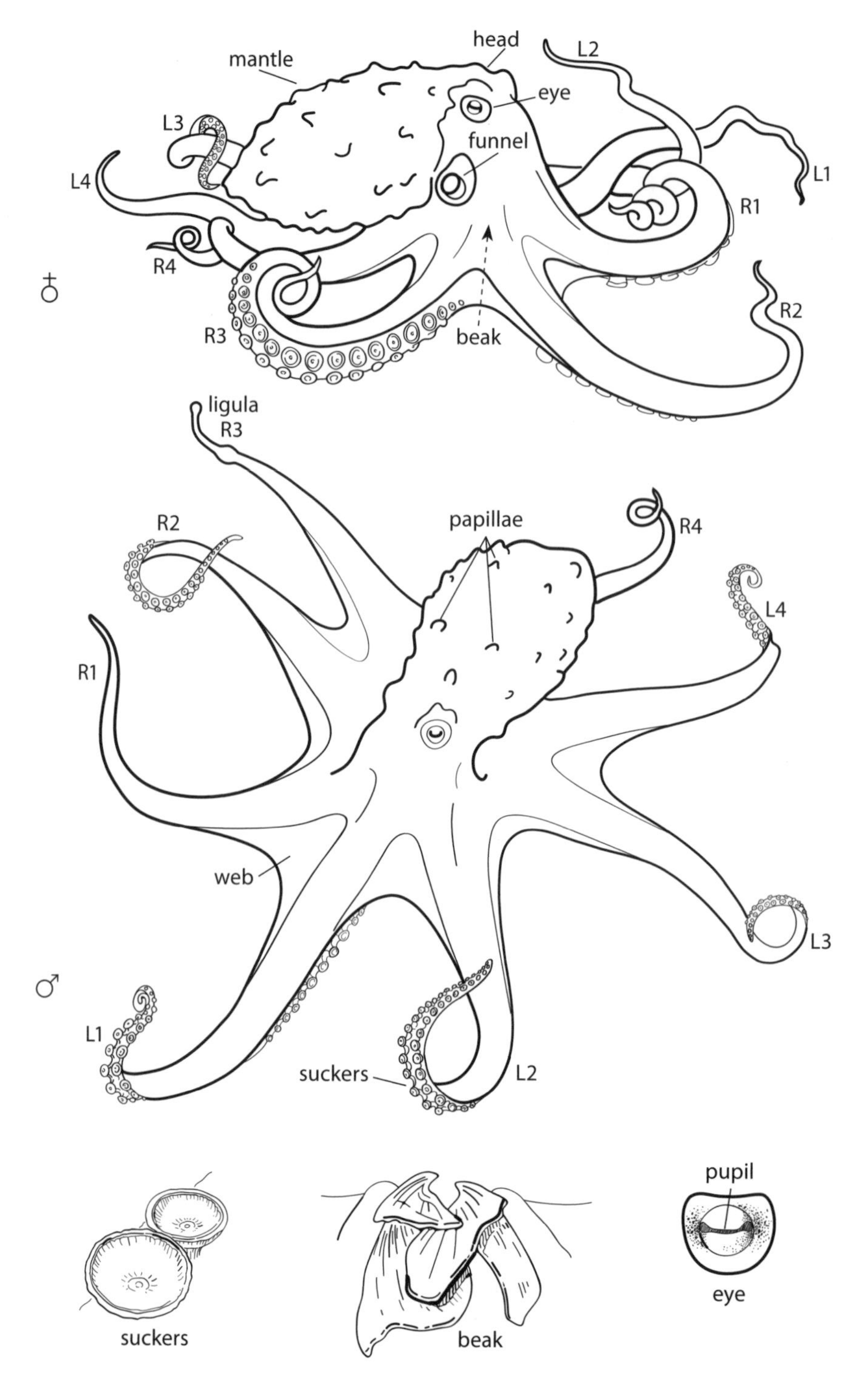

External anatomy features of female (above) and male (below) giant Pacific octopus (*Enteroctopus dofleini*). Marjorie C. Leggitt.

the gills and delivered oxygen to the blood, and it expels body wastes in its stream.

Because of this body arrangement, these marine animals are classified as cephalopods ("head-foots" in Latin) and named octopuses because of their eight arms (octopus means "eight-footed"), and they are also bilaterally symmetrical, with left and right sides mirroring each other, like in humans. The third right arm of adult male octopuses is modified to pass spermatophores, or sperm packets, to the female. Because the arms are lined with suckers along the underside, octopuses can grasp anything. And since the animal has no skeleton, it can flex its arms and move them in any direction. The arms aren't tentacles: tentacles are used for prey capture in squid, and these arms, with their flexibility, are used for many different actions.

Octopuses are found in all oceans at every depth and in many different marine habitats. Although we don't know exactly how many species of octopus exist, we think there are approximately 100 members of the genus *Octopus,* and probably about 300 octopods (members of the order Octopoda) around the world, and some do not yet have common names. Names aren't the only thing missing in the puzzle about octopuses: scientists are still piecing together their behavior, ecology, and physiology. We have seen juvenile common octopuses (*Octopus vulgaris*) in shallow tide pools in Bermuda and large giant Pacific octopuses (*Enteroctopus dofleini*) crawling out of the water to get over a rock jetty on the North Pacific coast. Some, such as the spoon-arm octopus (*Bathypolypus arcticus*), are adapted to the ocean's dark, cold depths up to 3000 ft. (1000 m) below the surface and can be found on the abyssal plain stretching across the bottom of the sea. Hawaiian day octopuses (*Octopus cyanea*) are found on shallow coral reefs, and argonauts (*Argonauta argo*) drift in the open ocean. The long-armed *Abdopus aculeatus,* which has no common name because it is so little known, crawls through near-shore sea grass beds in the eastern Pacific. The specialized deep-sea vent octopus (*Vulcanoctopus hydrothermalis*) is found, as its name suggests, near deep-sea hydrothermal vents way down at 6000 ft. (2000 m).

Octopuses will eat almost any animal that can't get away from them. Their menu depends on size. Tiny pygmy octopuses eat small hermit crabs, and adult giant Pacific octopuses favor big Dungeness crabs and geoduck clams. They have lots of ways of getting into shelled prey—scraping with their teeth, or radula, grasping with their parrotlike beak, even drilling a hole in the shell of a clam and injecting venom to weaken the muscles so the shell gapes open. The octopus's toxic venom also contains enzymes that start digesting the food. We found that the common octopuses we studied in Bonaire ate seventy-five different animal species. Sifting through the shells of their prey in the midden, or garbage heap, outside their sheltering homes, we discovered that their food ranged from immature conch snails and pen shell clams to shore crabs and shrimp. Some octopuses seem to specialize in one or a few species: the spoon-arm octopus eats brittle stars, one of the few animals common in the deep. Some octopuses have favorite species: the Hawaiian day octopus eats crabs almost exclusively, while the night-active white-striped octopus (*Octopus ornatus*) in the same islands eats cowry sea snails. All common octopuses adapt, as studies

Surfing the Web for Science

In 2008, Roland completed the data analysis for our study of whether octopuses can remember and recognize individual humans (yes, they can). One measure that distinguished octopuses' reactions to a person who fed them from their reactions to a person who hassled them by touching them with a piece of Astroturf-covered plastic pipe was a specific skin display, the eyebar. The eyebar extends the horizontal line of the octopus's horizontal pupil slit onto the skin on either side of the eye, presumably making the eye look less like an eye. An eye is a dead giveaway that you're facing an animal and not just a piece of the landscape. Many animals manipulate the appearance of the eyes, disguising their real ones or adding fake eyes as displays on the skin, to startle or confuse predators.

All we knew about the eyebar display at that point was that we had seen it when the octopus was annoyed. So I promised Roland that I would look into the structure of the display and its occurrence in different octopus species. How to begin? One situation that qualifies as annoyance to an octopus is being stalked by an underwater photographer, who appears literally in the animal's face, with lots of equipment and trailing streams of bubbles, letting off flash after flash in the octopus's eyes. It's not surprising that when I looked at pictures of octopuses, I saw a large proportion with eyebars.

across their huge range have shown, to eating whatever is most easily available where they live.

Even if they get very big, like the giant Pacific octopus, all the group live short lives. The longest octopus lifespan is three to four years, and most of the smaller octopuses die after about six months to a year. Most species start off very tiny, floating in and eating the tiny organisms of the plankton of the upper open ocean, and then they settle to the sea bottom as they get bigger. Young octopuses have only two things on their mind, eating and not getting eaten. They convert food to body muscle and fat very efficiently, partly because they are poikilotherms—they don't spend energy keeping their body warm. At the end of their life cycle, after being solitary animals, octopuses get interested in sex and find each other for mating.

I gathered photographs of octopuses from colleagues, especially Roland, James, and former graduate student Tatiana Leite, primarily of the common octopus and the giant Pacific octopus. But wanting more, I went to the Web. Finding countless pictures of live octopuses, I selected ones with eyebar displays to enter into my database. I chose the common octopus, the giant Pacific octopus, and the Hawaiian day octopus—species that live in different areas of the world, so that even if the Internet photo description didn't identify the species, I could figure it out from the location where the photo was taken.

After Tatiana ran the photos through a sophisticated data analysis, we found some intriguing things about the display. It's nearly always symmetrical—the line extends on the skin both forward and backward from the eye—and it's painted on the same mottled, pale or red background around the eye. The display is also different among the three species, dark and pencil thin for common octopuses, and wider, delineated by thin white stripes in the other two species. We don't yet know what situations cause octopuses to put the eyebar on their skin, but we do know what it looks like, which is a first step.

This was the first (and probably only) time I've done a research study without leaving my desk.

—Jennifer A. Mather

After mating, males become senescent and soon die. The female lays from fifty to tens of thousands of eggs, and tends them faithfully, keeping them well oxygenated, clean, and protected from predators. After the eggs hatch, the female dies.

Many octopus babies travel a long distance between the time they hatch and the time they settle to the bottom. Eggs of the smaller octopuses are only about 1/10 in. (3 mm) long. When they hatch, the babies rise to the surface of the water and are washed out to sea or along the shore, floating great distances on marine currents. If they survive, young octopuses eventually get heavy enough that they settle to the bottom, find a likely rocky ledge, muddy bottom, or boulder field, and make a home there. But this isn't true for all octopuses. In California, there are two species of two-spot octopus that look very much alike. One species, Verrill's two-spot octopus (*Octopus bimaculatus*), lays 20,000 tiny eggs that hatch into planktonic (open-ocean surface) young. But the second species, the Californian two-spot octopus (*O. bimaculoides*), lays 800 eggs that are five times as long as those of the other species, and when the babies hatch, they are benthic young—they just crawl out and start their life on the bottom of the ocean right away.

Divers and aquarium visitors are fascinated by the octopus's appearance. Octopuses can change what they look like in less than 30 milliseconds by expanding tiny pigment sacs in the skin, chromatophores. They can go from dark to pale, plain to patterned, rough to smooth, and a clumped shape to an elongated form. They change their appearance mostly to hide from predators, camouflaging to match the colors and the patterns of their background. Over brown algae, they will go blotchy brown, on sand they can change to tiny dappled grays and blacks, and under the shade of a rock they can become plum-purple. They can make half of the body pale and the other half dark (see plate 2). And when an octopus is finished matching its background and lifts off to swim away, it can put on a striped pattern, making it harder for a predator to track it. If all else fails, an octopus can squirt out a cloud of ink so the predator loses sight of it.

Although the octopus hides as much as possible to avoid trouble, there is another side to these animals: they can be deadly predators. They can bite into prey with their beak and inject venom from the salivary gland. A small Caribbean pygmy octopus (*Octopus joubini/mercatoris*) can catch a crab its own weight, and after one quick bite the crab stops moving and in thirty seconds it is dead. One of the most deadly marine animals is the

What's in a Name?

Throughout this book and in scientific publications, animals are referred to by their scientific names—by genus and species—though we also use common names when they exist. Why not just use common names? The answer is that they are not common, but vary with the location. Think of the robin, quite a different bird in England than in North America, but known by the same common name. In the early eighteenth century, Swedish scientist Carl Linnaeus developed the system of nomenclature for animals and plants, using ancient Latin and Greek linguistics to form binomial scientific names, and it has proved a great blessing for sorting out the diversity of animals and plants. For instance, wherever it occurs, *Octopus vulgaris* is an animal name everyone can understand. And the sorting also tells us about relatedness: every species in the genus *Octopus* is related to others in that genus, and is a bit more distantly related to those in the genus *Enteroctopus*, like *Enteroctopus dofleini*.

How do animals get their scientific names? The scientist who reports a species as a new one has the privilege of naming it or of having it eventually named after him or her. Sometimes an animal is named for a particular feature, like the two-spot octopus, *Octopus bimaculoides*. Sometimes the name refers to where the animal lives; a good example is *Vulcanoctopus infernalis*, which lives around deep-sea volcanic vents. Sometimes the name honors a previous researcher, so *Octopus joubini* is named after French naturalist Louis Joubin, who studied cephalopods in the nineteenth century (*O. joubini* was split into *joubini* and *mercatoris*). Sometimes the name is more whimsical, like *Wunderpus photogenicus*, which was known to divers as very photogenic years before it was described by scientists.

Just because scientific names are useful and commonly used doesn't mean they are fixed in stone; in fact they are fairly changeable. A species is an interbreeding group of animals that doesn't interbreed with other groups, and this can change. Sometimes these taxonomic decisions are challenging for those who are just studying behavior, since it makes us question what species we are working with. I've studied six octopuses over thirty years, and during that time, taxonomists have changed the species name of one of the animals I studied, the genus of another, and we are still not sure of the taxonomic relationships of *Octopus vulgaris*.

—Jennifer A. Mather

A Guide to the Mollusks in the Book

Here are genus and species names for the mollusks discussed in this book, alphabetically within each group, with common names when they exist, based on Mark Norman 2000.

Genus *Octopus*

Octopus abaculus	———
Octopus alpheus	Capricorn night octopus
Octopus bimaculatus	Verrill's two-spot octopus
Octopus bimaculoides	Californian two-spot octopus
Octopus briareus	Caribbean reef octopus
Octopus chierchiae	———
Octopus cyanea	Hawaiian day octopus
Octopus defilippi	Atlantic long-arm octopus
Octopus dierythraeus	red-spot night octopus
Octopus digueti	Diguet's pygmy octopus
Octopus insularis	———
Octopus joubini/mercatoris	Caribbean pygmy octopus
Octopus macropus	white-spotted night octopus
Octopus mimus	———
Octopus ornatus	white-striped octopus
Octopus rubescens	red octopus
Octopus vulgaris	common octopus
Octopus wolfi	star-sucker pygmy octopus
Octopus zonatus	———

Other Members of the Order Octopoda

Abdopus aculeatus	———
Ameloctopus litoralis	banded string-arm octopus
Argonauta argo	argonaut
Bathypolypus arcticus	spoon-arm octopus
Cirrothauma murrayi	blind cirrate octopus
Enteroctopus (Octopus) dofleini	giant Pacific octopus
Grimpella thaumastocheir	velvet octopus
Hapalochlaena maculosa	blue-ringed octopus
Ocythoe tuberculata	football octopus
Opisthoteuthis californiana	flapjack devilfish

Thaumoctopus mimicus	mimic octopus
Vitreledonella richardi	glass octopus
Vulcanoctopus hydrothermalis	deep-sea vent octopus
Wunderpus photogenicus	wunderpus

Other Members of the Class Cephalopoda

Architeuthis dux	giant squid
Cranchia scabra	glass squid
Dosidicus gigas	Humboldt squid
Euprymna scolopes	Hawaiian bobtail squid
Euprymna tasmanica	southern bobtail squid
Heteroteuthis dispar	———
Idiosepius pygmaeus	pygmy squid
Metasepia pfefferi	flamboyant cuttlefish
Nautilus spp.	nautilus
Rossia pacifica	stubby squid
Sepia apama	giant cuttlefish
Sepia latimanus	broadclub cuttlefish
Sepia officinalis	common cuttlefish
Sepioteuthis sepioidea	Caribbean reef squid
Vampyroteuthis infernalis	vampire squid

Other Members of the Phylum Mollusca

Aplysia californica	sea hare
Arca zebra	zebra mussel
Archidoris montereyensis	sea lemon nudibranch
Clione limacina	sea angel
Crepidula fornicata	slipper limpet
Ctenoides spp.	file clam
Humilaria kennerleyi	Kennerley's venus clam
Janthina janthina	janthina
Lima spp.	file clam
Melibe leonina	lion nudibranch
Pinna carnea	pen shell clam
Protothaca staminea	Pacific littleneck clam
Strombus gigas	queen conch
Tegula funebralis	top snail
Venerupis philippinarum	Manila clam

blue-ringed octopus (*Hapalochlaena maculosa*) of Australia; its venom can be fatal to humans (see plate 3).

Under this variable skin, an octopus, like other mollusks, has several ganglia, collections of nerve cells that help to regulate activity, spaced around their body. In the octopus, the ganglia are centralized in the brain, which is a large one for invertebrates. Besides these nerve centers, a huge part of the octopus's brain is devoted to controlling the skin display system, and two areas are devoted to storing learned information. We tend to think that nerve cells belong in the brain, but in the octopus three-fifths of the nerve cells are in the arms. They are probably needed there to control the very complicated movement that octopus arms and suckers can make.

The blood of octopuses and other mollusks is blue because of their oxygen-carrying pigment, hemocyanin. Hemocyanin isn't as efficient as our hemoglobin at carrying oxygen. For the slow-moving standard mollusks, that doesn't matter much, but since the octopus's physiology is faster, its circulatory system is modified. The blood is circulated in a closed system, a series of arteries going out to the body and then veins bringing the blood back to the heart, like in humans. Still, this system isn't adequate for the octopus, so the animal has evolved two accessory hearts at the base of the gills that push blood out to get it to the oxygen faster.

Mollusks are in the phylum Mollusca. Since the head-foot group is so different from the snails and clams, they are defined as the class Cephalopoda. Among the cephalopods, there's one really different group, the few species of *Nautilus*, which are remnants of a huge group that dominated the seas in ancient times; they are slow, live in deep water, and have a scavenging lifestyle. The rest of the cephalopods are the subclass Coleoidea, comprising four orders: true squid, or Teuthoidea; cuttlefish, or Sepioidea and Sepiolida; one deep-sea species of vampire squid, or Vampyromorpha; and the octopuses, or Octopoda. Animals in these four orders have the same basic body plan, though the addition of two elastic tentacles defines the squid as decapods (or ten-footed).

Octopuses face ordinary challenges in unusual ways. Their amazing mobility lets them go almost anywhere they want on the sea bottom to find a hidden crab or snail. The changeable skin allows them to conceal themselves against a wide range of backgrounds or to dazzle a potential mate. They can walk, swim, or pull themselves along the bottom with extraordinary grace, and they can solve problems that other mollusks might not even comprehend. All of these characteristics make octopuses fascinating to study.

In the following chapters, we take you through the life of an octopus. We start with the octopus egg, in chapter 1. Then we describe the issues that matter to an octopus as it goes through the different stages of its existence, and we conclude with mating and the end of life. In the final chapter, we discuss other cephalopods related to octopuses that help us understand the octopus. Then we offer a guide to obtaining and keeping an octopus in an aquarium.

In this book, as authors we refer to each other as "we" or by our first names. Other research findings are cited by first and last author names, and those studies as well as ours are included in the reference list at the back of the book. In the text, you'll find that sometimes the squid is used as the basis of discussion when no good octopus example is known, because squid are similar in structure and physiology.

By the way, the plural of octopus isn't octopi, because the word is Greek—*octopous* to be exact—not Latin. The Greek plural would be *octopodes,* but we call them octopuses.

1

In the Egg

Octopuses are oviparous: they lay eggs like chickens do. Some animals are viviparous, like humans: offspring grow inside the mature female, who then gives birth to fully developed young. Other animals are ovoviviparous, like snakes: the female essentially lays eggs inside herself, the eggs hatch inside, and she gives birth to live young. Almost all octopuses lay eggs outside their body and the embryos develop inside the eggs. The process of producing viable eggs and ensuring that the young successfully hatch out is the most important task of a female octopus. This task comes at the end of her life, and she totally devotes herself to it.

The female octopus has two ovaries, where the eggs are produced. As in many other invertebrates or fish, the octopus's ovaries are cream-colored and granular. The ovaries of mature females take up much of the space inside the body: the egg masses may be 25 to 30 percent of the female's weight in some species, like the red octopus (*Octopus rubescens*), and up to 40 percent in other species. If humans had this high a gonad index, a 120-lb. (55 kg) woman would have a 48-lb. (22-kg) baby!

During mating, the male inserts a spermatophore, or sperm packet, by passing it along its arm to the oviduct of the female. There, the spermatophore turns inside out, the sperm are pushed to the end of the packet nearest the oviduct, and the end of the spermatophore bursts open, releasing the sperm. The sperm are stored in the wall of the oviduct in a special organ, the spermatheca, until they are needed to fertilize eggs. When eggs travel down the oviduct before they are laid, they pass through the spermatheca and get fertilized. We know that live sperm can be stored in the spermatheca for a long time, because, for example, a female giant Pacific octopus was shipped from Tacoma, Washington, to an aquarium in New York City. She was kept alone for seven months and then laid fertile eggs.

Before the female octopus can lay her eggs, she must find a suitable place for them. Most shallow-water octopuses, like common octopuses and

Hawaiian day octopuses, lay their eggs in a den. The den might be in a crevice or cave in a rock wall, in an excavation under a rock, or under human-made underwater structures such as concrete blocks, pilings, or shipwrecks. Red octopuses have laid eggs in beer bottles, aluminum soft drink cans, and even in an old shoe. Smaller species of octopuses, such as the Caribbean pygmy octopus, may lay their eggs inside clamshells or even a beer can. And several octopus species carry their eggs with them, not laying them in dens at all. The female argonaut, a pelagic octopus, makes a delicate and minimally coiled shell up to 18 in. (46 cm) across in which to lay her eggs.

If undisturbed, octopus females almost never leave their clutch of eggs. For better protection, a number of shallow-water female octopuses even block up their den openings with rocks, shells, or other material, and don't move the obstructions until the eggs are about to hatch. Hidden dens make it difficult for scientists to find nesting female octopuses and study this important part of their life cycle; Jim Cosgrove (1993), because he has such intimate knowledge of their habitat, has successfully found the dens of nesting giant Pacific octopuses.

Like a chicken's egg, what we call an octopus egg is a complete package, with cushioning material, a yolk for nutrition, and a shell for protection. As a fertilized octopus egg passes down the oviduct getting ready to be laid, it goes through the nidamental gland, where it is coated with a nutritious and protective, clear jelly. It then goes through the oviducal gland, which forms a protective sheath, the chorion, around the egg and its jelly. In some octopus species, the egg is then ready to be laid and guarded by the female. Octopus eggs are typically oval or teardrop shaped, and most are tiny, the size of a grain of rice.

In most octopus species, the chorion is drawn out on one end of the egg to form a stalk. In some species, the female lays large eggs and uses this stalk to attach each egg singly to the ceiling of her den. In other species where the female lays small eggs, a long strand or string is produced by the oviducal gland, and strings of many eggs, up to several hundred, are delicately woven together by the suckers on the female's arms near the mouth to form a festoon of small eggs. Each festoon is then attached individually to the ceiling of the den, forming a cluster. In some species, such as the common octopus or the giant Pacific octopus, there may be hundreds of festoons with tens of thousands of eggs. The female may take more than a month to deposit all these eggs in the den.

The Mysterious Argonaut

Argonauts (*Argonauta argo*) are an enigmatic group of octopuses that are highly specialized. Females begin to build a coiled paper-thin shell like a cornucopia when they are young, and they live in it and lay their eggs in it. They secrete the shell with greatly modified paddle-shaped arms. These arms cover the shell, even when they are swimming. The shell doesn't provide any flotation and these octopuses are good swimmers, living in the open ocean in tropical and subtropical zones, swimming constantly with their water jets to keep them from sinking.

The argonaut was named after the crewmen of the ancient Greek mythological hero Jason, who made epic journeys aboard the ship *Argo*, because argonauts were found "wandering" in all the tropical oceans. The coiled argonaut shell looks somewhat like a nautilus shell, but argonauts are different from chambered nautiluses (*Nautilus* spp.): the shell is part of the body of nautiluses, but female argonauts can leave their shell and then reenter it.

Very little is known about argonauts. We know the eggs are fertilized somehow by pygmy males about an inch long (2.5 cm), which have never been seen in the wild. The males somehow deposit a detached arm containing sperm into a female, and she then uses the sperm to fertilize the eggs. We know little about what argonauts eat, although ones in aquariums have eaten shrimp and small fish. No one has kept them in aquariums longer than about three weeks. Young coming from the tiny eggs, just 1/10 of an inch (2 mm), have never been raised. There is also some indication that argonauts are social. Chains of six to eight argonauts have been photographed, each female holding onto another's shell. We don't know if this was a chance occurrence or deliberate.

We also know little about species of argonauts. They have been identified by characteristics of their shells, largely from those that have washed up on shore, not the actual animal. There appear to be six or eight species with several more undescribed. Norman (2000) tells of a mass stranding he investigated that took place in Australia. Thousand of females with eggs in their shells washed up on the beach, and most of the females and their eggs died. Such mass strandings of argonauts are not uncommon; they may occur from weather or oceanographic conditions.

—Roland C. Anderson

In general, octopuses lay either tens of thousands of tiny eggs or approximately 100 larger eggs, depending on the species. The eggs of the common octopus are about $1/8$ in. (3 mm) long, while those of some deep-sea species are more than $1\frac{1}{2}$ in. (40 mm) in length. Each strategy works, or else the species would die off.

Scientists have described the different reproductive strategies of animals in terms of their ecology and life style as a continuum on a scale from r to K. The rate of population increase is the r factor at one end of the scale, and K stands for the carrying capacity of the environment at the other end. Animals using the r strategy mature early at a small size, have a short life span, produce a large number of young with no parental care, and die shortly after reproducing. Animals using the K strategy mature later in life, live longer, produce fewer young with parental care, and can reproduce more than once. Different octopus species fall different places on the r–K line. The r strategy is used by those species producing planktonic young, and those producing benthic young are farther toward the K end of the scale. The deep-sea spoon-arm octopus is one with the highest K value; it produces just a few young and has a relatively long life span of four years or more.

Some species of closely related octopuses are known as sibling species. They appear alike because of their physical features, genetic makeup (or cladistical comparison), and they may even live in the same areas. There are sibling species of zebra-striped octopuses (*Octopus chierchiae* and *O. zonatus*) on each side of Panama that were presumably the same species at some time before the isthmus arose. Some of these sibling species have a different place on the r–K line, which also distinguishes them from each other. Sometimes you have to wait until the female lays eggs to know which species of octopus your individual is (see plate 4).

When female octopuses, such as the Caribbean reef octopus (*Octopus briareus*), put their resources into laying larger eggs, the eggs hatch out larger juveniles that are better prepared to take on the benthic world than their smaller relatives. These larger juveniles are able to crawl, hide, change color, ink, or swim from predators. In other words, they assume a normal subadult octopus life.

We humans use the K strategy. We give birth to just one or a few large offspring, although they are certainly not ready to live by themselves at birth. Anthropocentrically, we may think that having a few large eggs, large offspring, and parental care comprise a good strategy, perhaps the best.

This strategy in octopuses produces juveniles that are well able to live an adult life style on the ocean's bottom. But in octopuses, such a strategy prevents a wide dispersion of the species, because they can't get very far from where they grow up: they must crawl or slowly swim to colonize new areas. So they tend to have small ranges, and a species in a small area can be threatened or even wiped out by ecological disasters, such as storms, El Niño currents, or pollution. Examples of this kind of event are the endemic land snails of Hawaii, which have gone extinct from being eaten by another introduced snail; the many freshwater mussels of the Ohio River that have gone extinct from pollution; and the threatened snails in Banff, Canada, that only live in three hot springs. Because of their lack of dispersion, more freshwater and land mollusks have become extinct than all mammals and birds combined.

The other strategy, producing many small eggs and tiny hatchlings, also has its advantages and disadvantages. A primary disadvantage of the production of small hatchlings is that most of them get eaten, although selection means that the fittest survive. Tiny octopus hatchlings look somewhat like miniature octopuses but have stubby arms, an attached yolk sac, and a few chromatophores. They may or may not have inking ability yet. They swim and drift in the plankton until they are large enough to settle to the bottom and take up a usual octopod benthic existence. These planktonic hatchlings are called paralarvae: they don't go through a true larval stage (looking quite unlike the adult), like caterpillars do before becoming butterflies.

In addition to having few resources when they hatch and needing to eat soon, the planktonic paralarvae are likely to be eaten by larger animals that eat plankton. But since there may be thousands of these hatchlings, the chances are good that a few will survive. And since they are planktonic, they are often carried by currents to new areas of settlement. Generally, the more eggs produced by an octopus species, the larger its range, since the paralarvae can float in currents for several months. Production of many eggs may also be good for the gene pool of the species, because it increases the odds of having a variant that might increase survivability.

As far as we know, all female octopuses that lay eggs guard them. They use several methods of protecting the eggs, depending on the species. The football octopus (*Ocythoe tuberculata*) is the only known octopus that is ovoviviparous: the fertilized eggs are retained in the extended oviducts until they hatch, and the hatchlings are released directly into the water.

Since the female argonaut is an open-ocean drifter, she can't use rocks for protection. She secretes a thin calcareous coiled shell with the webs of her first arm pair, inside which she attaches her eggs. She then sits in the shell, protects the eggs from overgrowth by algae, and keeps them clean and oxygenated. She lays the smallest known octopus eggs, just less than 1/10 in. (2 mm) long.

Some deep-sea octopuses have a bigger problem. On the abyssal plain of the deep ocean, there is a thick layer of mud and few hard things to attach eggs to. This mud results from the sediment, plankton, and feces that drift down through the water's layers. So octopuses on the bottom attach their eggs to anything hard they can find: a rock outcrop, a whalebone, a shell, or a shipwreck. A sheet of plastic trawled from the bottom of the North Atlantic had eleven female deep-sea spoon-arm octopuses guarding a total of more than 100 eggs attached to it.

Octopuses give no care to their hatchlings; the females die about the time their eggs hatch. Once the eggs hatch, the paralarvae, or juveniles, are on their own. But females are wonderful at guarding their eggs. They usually find a den that is protected and has good water conditions—high oxygen, low pollutants, and medium water currents. But even members of the same octopus species have different personalities, so this and chance lead them to establish their maternal dens in different areas, sometimes in places that are not very safe.

While the mother octopus tends her eggs, she stops them from getting fouled by overgrowth of marine algae and settling organisms such as hydroids, barnacles, and tunicates that might grow on the eggs (see plate 5). She does this partly by constantly blowing water through her funnel along and between the eggs or egg strands. The festoons are actively moved and bathed by her water jets, so they must be attached securely in order to take this treatment twenty-four hours a day for the about six months it takes for a giant Pacific octopus egg to hatch. She also manipulates the eggs with her arms, grooming them with her suckers and arm tips, which go snaking through the eggs to remove any fungus or algae growth on eggs that might choke or kill them. The egg strings are made of a chitinous, or fingernail-like, material, to hopefully be tough enough to withstand this constant manipulation.

Because the female octopus usually doesn't eat while she is guarding eggs, she may lose up to 50 percent of her body weight during development of the eggs. Not eating while brooding the eggs has several advantages. First, she doesn't foul the den with food wastes or feces, which helps ensure good water quality for the eggs. She certainly could leave the den to find food or eliminate body wastes, but that would mean exposing her eggs and herself to danger from predators. Fish of several species follow foraging octopuses, hoping to snatch a bit of food, and being out eating would advertise a female's whereabouts. Second, she won't produce a midden of shells or other food remains in front of her maternal den. Some octopus predators (including humans) target den middens, finding them by sight or chemical cues. Third, not eating eliminates any chemicals arising from her metabolism or in her feces. Moray eels can find octopuses in their dens, but it is not known yet what chemicals from octopuses they sense. It is possible they sense some product of food metabolism that is not present while female octopuses are guarding eggs, rather than the body tissue metabolism females undergo while brooding.

While she is tending her eggs, the female octopus survives by metabolizing muscle tissues (octopuses don't use fat for metabolism as we do, and have very little fatty tissue), so she deteriorates considerably at the end of her life. She turns gray or pale, as though she can't change color any more.

Deadly Dedication

In Washington state's Hood Canal, there are several sites known for their giant Pacific octopuses, with rocky outcrops having many crevices suitable for dens. Unfortunately, the waters of Hood Canal experience a period of low oxygen each fall when phytoplankton and macroalgae die, decompose, and use up oxygen in the decomposition process. Oxygen in the water during these periods has been recorded as low as 2 parts per million (ppm), compared to a normal 7 or 9 ppm. Fish and octopuses leave the deep oxygen-starved water and move into the shallows, which have more oxygen for the duration of this event. But dedicated female octopuses guarding eggs can die at the nesting sites, along with their eggs, since they won't leave the eggs. Octopuses move back into the deep water a month or so after this occurrence, either down from the shallows or migrating from other areas.

—Roland C. Anderson

She hardly moves. She shrinks to half her size, and actually looks old and wrinkled. She can open her den if she has blocked it up, and may manipulate the eggs somehow to stimulate them to hatch. If she is guarding small eggs, she will blow the paralarvae out of the den, usually at night, giving them a boost into open water.

Sometimes females live after their eggs hatch and go into a state of senescence, but this behavior is more commonly seen in males. Senescent octopuses don't hide in a den, but crawl haphazardly over the sea's bottom, unconcerned about predators or prey. Evolution sets a fine balance between the survival of the female and the survival of the eggs; most females live until after the eggs hatch, since dying sooner would be a disaster for the potential progeny.

The egg laying, guarding, and grooming processes have become well-shown by Olive the Octopus, a giant Pacific octopus who laid her eggs at a popular dive site in Seattle's Puget Sound just offshore from the busy downtown area in 2002 (see plate 6). Olive was first seen guarding eggs in Cove 2 near Armeni Park. Divers guessed she weighed approximately 60 lb. (27 kg), based on the size of her largest suckers. A nearby large male octopus had been named Popeye, after the cartoon character, so she was named Olive, after Popeye's girlfriend Olive Oyl. A group of divers dove at the site every Tuesday night and reported their findings to us at the Seattle Aquarium.

Olive made her den under a cluster of sunken wooden pilings called a dolphin. The dive site is an area of ongoing boating activity, and the dolphin may have been tipped over and sunk by a storm or rammed by a boat. The four pilings are bundled together by steel cables and lie on the bottom in 100 ft. (30 m) of water, parallel to shore. She made her den under this dolphin, midway along it, with two openings, one shoreward and one facing toward deep water. Before laying her eggs, she looked out of the deeper opening through the cool 50°F (10°C) water.

There was little evidence of food remains in front of either den opening at the start of her brooding. We have found that the normally hard shells of red rock crabs, a common prey, become thin and fragile within a few days, and those near her den were hard, so she had either just stopped eating or she was reusing a den recently occupied by another octopus. The cluster of sunken pilings had several dens under it along its length. Unlike most other shallow-water octopuses, she did not wall up the entrance to her den with rocks. Instead, she created a fence of 8-in. (20-cm) rocks in a semicircle in front of the deeper opening, and she didn't put anything in front of the shoreward one. During her entire brooding period, she was highly visible to hundreds of divers.

Divers first saw her eggs on February 25, 2002. She laid the characteristic strings of eggs on the ceiling of her den, attached to the underside of the wooden pilings. That day, she was observed in an upside-down posture with her suckers facing upward, so it is likely she was still in the process of laying eggs. No one counted the eggs, but giant Pacific octopuses characteristically lay about 70,000 eggs. Larger females lay more eggs and smaller ones lay fewer. Olive was a bit larger than normal, so she may have laid about 100,000 eggs.

On later dives in following weeks, divers saw her right side up, blowing water through the eggs and caressing them with her arm tips. At this time, she was normally a dull gray color, but she turned red-brown in response to divers' bright underwater lights or a gentle touch. During the month after the eggs were seen, she would take a piece of herring offered as food by divers, but later she wouldn't eat, and blew offered food assertively out of the den.

She guarded her eggs through the summer, seemingly unfazed by the hundreds of divers viewing her. She pushed sunflower sea stars away from her brood chamber and fended off other egg predators that hovered nearby. She ignored octopuses that made short-term dens nearby under the dolphin, even when they mated as close as 50 ft. (15 m) away. She probably didn't notice their absence as they moved away, the female to make another maternal den of her own somewhere else and the male going off to die.

During that summer, Olive behaved normally for a giant Pacific octopus guarding eggs: she refused food, she was never seen out of her den, she constantly kept the eggs clean, she repelled predators and egg eaters, and she grew unresponsive to divers, maintaining a gray color that gradually turned to a translucent white. Her eggs were white when first laid, but gradually changed to a yellow color as the embryos grew within, and then turned brown with chromatophores just before hatching. Divers saw eyespots inside the eggs in mid June, about 110 days after the eggs were laid, so they knew the eggs were fertile.

Divers witnessed the first of Olive's eggs hatching on a night dive on September 22, 209 days after the eggs were laid. While a few of the paralarvae swam out of the den during the daytime, Olive blew most of them out of the den at night. She may have been causing them to hatch at night by blowing strong water jets over them. At this point, she was totally white, almost translucent. Her skin had several large white ulcers on her arms and mantle, and it had the appearance of rotting away. She was totally unresponsive to divers, even when touched, devoting her remaining energy to her eggs.

The peak of the paralarval hatching was October 7, 224 days after she laid her eggs. Divers saw the last hatchlings on October 31, and at that time there were virtually no eggs left unhatched. Olive was dead on November 6, 254 days after laying her first eggs. Her body was about 6 ft. (2 m) from the opening of her den, being fed on by two sea stars. Nothing goes to waste in the sea, and scavengers are always waiting for the chance to clean up dead bodies.

The precise timing of her death to the last hatching of her eggs is remarkable. Although it sounds anthropocentric, it looked as though she clung to life until she knew her eggs hatched. It is also remarkable that she was able to bring her eggs to successful hatching considering her circumstances. Her den was located within Seattle's inner harbor, next to the outlet of a river that flows through an industrial area. The river and the harbor were once very polluted. Her success may be a testament to our modern pollution cleanup efforts and awareness of the necessity of keeping pollutants out of rivers and bays, or it may simply reflect the durability of this female and her eggs.

Olive was visited almost daily by curious divers. Her home was Washington state's most popular dive site, used by many dive classes as their first open-water dive each week, since the location is sheltered from storms and their waves. Her success despite all these disturbances is also a testimonial to her dedication to the eggs.

The saga of Olive the Octopus brooding her eggs in Seattle's harbor was covered by several local newspapers and magazines. The reading public was entranced with her story and saddened by her death. The diving community also mourned her death. One dive magazine ran an article about her, "So Long, Olive, We Barely Knew You," lamenting the short life span of octopuses.

Many octopus eggs take a long time to develop. While the 224 days Olive's eggs took to hatch is a bit more than the about six months reported for the species, this egg development period is by no means the longest for an octopus species. Egg development time is dependent on the temperature of the water: the colder the water, the longer the development period within the egg. Eggs of the giant Pacific octopus in California have a four-month development, while those in Seattle or Alaska may take seven to eight months to hatch. The spoon-arm octopus of the North Atlantic and the Arctic Ocean lives on the continental slope, in water that is 600 to 1200 ft. (200 to 400 m) deep and a temperature of about 35°F (2°C). James has raised this type of little octopus through several generations, and he has found that their eggs take over a year to hatch out into benthic juveniles, ready to take up the bottom-dwelling existence of their parents.

Most deep-water octopuses have large eggs with long development periods that hatch out looking like their parents. This is logical when you consider where they live. There is nothing small like plankton in the ocean depths for the paralarvae to eat, so they must be large enough to be able to eat larger organisms. Some moderately deep-water fish spawn eggs that float to the surface waters, where their hatchlings live in the plankton, only to swim deeper when they are older and large enough to undertake such a vertical migration. But they may only migrate down a thousand feet or so. Other deeper-living fish and octopuses don't use this risky strategy, nor do most other creatures that live a mile deep.

Biologists Janet Voight and Tony Grehan, filming a rocky outcrop rising above the muddy abyssal plain off the west coast of Canada from a submersible in 2000, made a fascinating discovery: they saw twenty-eight female octopuses (of unknown deep-water species) guarding eggs laid on the rocks. They had laid eggs in different places and each had laid fewer than 100 large eggs, attached singly to the rocks. Some of these eggs were the largest known of any octopus, about 2 in. (5 cm) long. After some of these eggs were collected, they hatched out into the largest octopus hatchlings known to science. One female was still guarding her eggs in the same location a year later, and she looked senescent the second year, much like Olive toward the end of her brooding period. Based on the rate of development of the eggs, their size, and the temperature of the water at those depths, we believe it may take up to four years for the eggs of this species to hatch, presumably guarded all the time by a fasting female octopus. At least the deep-water females don't have as many predators to watch for as the shallow-water species do, but the brood time is remarkable. The possibility that the female doesn't eat during such a long guarding period is something to think about, and extrapolation from the brooding time leads to a possible life span of that species of over ten years, the longest for any octopus. Everything's slower in the deep.

Researchers have collected developing octopus eggs and viewed them under light with low-power microscopes or opened them up to see several stages of the development that occurs inside the egg. Huge changes take place in the egg capsule, starting with a cell and ending up with a complete and fairly well-developed animal. Since the egg capsules are usually opaque, embryologists (who study development before birth) can sometimes see what's going on inside, but often they have to open up an egg, preserve the embryo, and study it later. The single fertilized egg cell divides into clumps of cells, arranged in blastula and then gastrula stages. Gradually these cell collections begin to specialize, and we can see the beginning of the adult organs. Meanwhile, the little embryo flips its position in the egg twice, ending up with its mantle pressed against the opposite end from the attachment, ready to push out in the world during hatching.

When does an egg become an octopus? The changes are gradual; there's no specific time, as in all embryological development. First, the yolk develops to nourish the embryo and can be seen extending toward the egg capsule attachment. Then the arms develop, first visible as eight arm buds around the yolk and gradually getting longer (though the planktonic octopuses like the giant Pacific octopus don't have very long arms at birth). The eyes begin to develop at this point, and since they are dark with their pigment, anxious aquarists and the divers visiting Olive could see them through the capsule and know the eggs had been fertilized. Next, the heart begins to develop well enough to be an organ and to beat. The final part of the octopus that can be seen to develop is the chromatophores—there are not many of them and they are conspicuously large in the semitransparent skin. All through the embryonic period, the yolk is used to nourish the embryo, and by the time of hatching, none is left. Sometimes an octopus is disturbed during her brooding, and she may inadvertently push the eggs around so they hatch early. If so, they will have remains of the yolk sac like a deflated balloon sticking out from between their arms.

Hatching must be a traumatic event in the life of the octopus embryo, just as being born is to humans. We go from being cushioned, warmed, protected, and nourished by our mothers inside their bodies to living by ourselves. We go from not having to breathe to having to expel liquid from our lungs and drawing our first breath. Through the course of vigorous muscular contractions, we are expelled out the birth canal into the external world, much like other mammal babies. Chickens and other birds have to chip their way out of their tough eggs. Frequently the chicks have a hatching tooth that helps them penetrate the eggshell, which is later resorbed or falls off as the chick grows up. Some reptile mothers such as crocodiles help the eggs hatch by gently cracking the eggs, and mother octopuses may also help the eggs hatch.

Octopus hatching occurs with the aid of a hatching gland, a collection of enzyme-containing cells on the mantle of the embryo that help dissolve the chorion (egg shell), along with violent expansions and contractions of the mantle. The little octopus paralarva breaks mantle-first through the distal end of the egg, popping out in the normal swimming posture of a jetting octopus. The octopus lives on a remnant of its yolk sac for the first few days, but that is soon absorbed, and the hatchling must find food for itself. Benthic hatchlings, like those of the Caribbean reef or Californian two-spot octopus, just crawl away, but paralarvae of species such as the red octopus take up the life style of a drifter, swimming in the rich surface waters of the sea.

2
Drifting and Settling

While most octopuses live on the sea bottom, the situation is different for the newly hatched young. The many species of octopuses have two different life style strategies as hatchlings. These differences come from the size of the eggs, not the size of the adult animals: the giant Pacific octopus is one of the biggest octopuses but has tiny young. Most octopus species lay small eggs that produce small baby octopuses that are planktonic—they get washed away by ocean currents and temporarily live in the plankton-rich surface waters of the ocean. A few species lay large eggs that produce benthic hatchlings, big enough to live on or near the bottom of the ocean.

Living in the plankton requires adaptations, especially for the early stages of an animal like the octopus that later spends its adult life crawling on the bottom of the ocean among the rocks or coral. To understand why many young octopuses start their lives in the plankton, we must understand what plankton is, how the plants and animals that comprise plankton live, what methods they use to keep from sinking, what they eat, how they swim, what preys on them, and how they avoid being eaten. We must understand the advantages and disadvantages of living in the plankton, the bigger ecological implications of having a life stage different from that of an adult, and the adaptations and changes needed to live that life stage in that environment.

The word "plankton" is derived from the Greek root word that means free floating or wandering. The term is applied to any plant or animal that is unattached and floating in the surface waters of the ocean as well as to any weak, swimming animal that cannot swim against the ocean's currents. Even some comparatively large creatures such as jellyfish and the open-ocean argonaut octopus are part of the plankton. Relatively huge creatures such as sea turtles and ocean sunfish, which are weak swimmers, are sometimes considered planktonic. Plankton includes marine bacteria, animals, and plants, and some swimming animals that have plant pigments that undergo photosynthesis (making food from carbon and oxygen), which is normally a plant trait.

In her poem *Plankton,* Joan Swift wrote:

> They live their lives unseen
> Not just gray mobs without faces
> But like calm, steady workers
> In some underground plot
> To keep the world alive.

Plankton does indeed keep the world alive and is pretty much unseen and under-appreciated.

Animals and plants living in the plankton are the most important organisms in the ocean, but they are neither the most visible, the best known, the most highly beloved, nor the most feared. They are not the whales, the sharks, or the sea stars. They are not the familiar animals we eat, such as clams or oysters. They are not the organisms we mostly study, like porpoises, or the animals we capture to exhibit in public aquariums. They are usually not octopuses, although some octopuses are part of plankton as juveniles and a few as adults. In *The Log from the Sea of Cortez* (1951), writer John Steinbeck said that the disappearance of plankton, although the components are microscopic, would in a short time probably eliminate every living thing in the sea and ultimately the whole of human life. Since he wrote that, we have learned of the ecological communities surrounding the deep-sea hydrothermal vents that are not dependent on plankton for their survival, but instead depend on the hydrogen sulfide there, and they use bacteria to break it down and gain energy. But everything else in the ocean depends on plankton.

Plankton can be divided into plants (phytoplankton) and animals (zooplankton). Most animal phyla have representatives in the plankton, at least at some stage of their lives, since many animals such as fish and octopuses spend their juvenile stages there. Those creatures that are only in plankton temporarily are known as meroplankton, such as octopus paralarvae, while those that spend their whole lives there are called holoplankton, such as small single-celled plants, or diatoms.

The animals and plants of the plankton are extremely important to the ecology of the oceans. First, the plankton provides food for most of the

Counting Copepods

I got an idea of how many plankton there are when I was an undergraduate at the University of Washington in Seattle. As part of the laboratory experience in a biological oceanography class, we had to separate a plankton sample into its constituent parts, to species level. The sample came from the Chukchi Sea, northwest of Alaska. It was almost all copepods, collected with a large plankton net towed behind the oceanographic research vessel *Thomas G. Thompson*.

The sample consisted of a quart-sized jar filled with microscopic organisms. I don't remember how long a plankton tow this represented or how much water was strained to get this amount of plankton. The plankton was divided into equal portions for each member of the class. The class was not large, maybe a dozen students, and we each had to separate our sample into six species. My sample was in a specimen bottle about the size of my middle finger and was filled with copepods. After several full afternoons poring over a microscope and breathing formalin, we ended up with smaller specimen jars, each with a precise number of different species.

This tedious chore gave us a valuable experience in the field of taxonomy, the science of zoology that divides animals into species. And the exercise gave the principal investigator on the staff an idea of water conditions where the sample was taken, since different species live in different conditions.

Copepods are the most numerous animals on Earth, and they are very important. Large planktivores such as whales eat them, as do juvenile octopuses drifting in the plankton the first few months of their lives.

—Roland C. Anderson

ocean's other animals. Phytoplankton provide food for zooplankton, which in turn are eaten by fish and by other tiny organisms like paralarval octopuses. Little fish are eaten by big fish, and fish are eaten by sea birds, squid, marine mammals, and humans. The world's largest fish, the whale shark, and the world's largest animal, the blue whale, both live on plankton. Dead plankton and their waste products drift down to provide food for mid-water animals and those living on the abyssal bottom. The numbers of animals and plants in the plankton are tremendous. Terrestrial insects represent the most species (over a million have been described), but there are more copepods, small weak-swimming crustaceans, in the ocean than any other animal on earth.

Second, a huge part of plankton is made up of plants or other organisms that have chlorophyll and produce their own food. As a side effect of this process, the oceans produce 50 percent of the oxygen in the earth's atmosphere, 95 percent of which comes from phytoplankton and 5 percent from bigger algae (kelp and seaweed) or the few marine vascular plants that live near shore. These organisms also provide habitat for marine animals, like mature octopuses and their prey. At the same time, the plankton absorbs carbon dioxide and uses it for metabolism. Because of the large volume and surface area of the oceans, two thirds of the earth, this activity provides an enormous buffer to the oxygen–carbon dioxide budget of the earth's atmosphere. Carbon dioxide is currently increasing in the earth's air despite this buffer, because we humans create huge amounts of it. Plants on land and in the sea can't keep up with our industrial emissions. Plankton may also lessen the atmospheric ozone layer that contributes to global warming and bleaches the coral habitats in which many octopuses live. Clearly, the plankton in the oceans is immeasurably important for our survival.

Third, plankton is important as an energy source. Dead plankton falls to the ocean bottom and collects there slowly, under ½ in. (1 cm) in 1000 years but amounting to thick deposits over millions of years. The rich oil deposits around the Gulf of Mexico are the result of plankton deposits on the floor of an ancient sea there millions of years ago. While it's a harsh environment, some specialized octopuses such as the spoon-arm octopus live there.

Fourth, both live and dead plankton can be ecological indicators. There are some plankton that exist only under specific climatic conditions and some that behave differently at different temperatures. By looking at fossil plankton, scientists can tell what the climate was in the past. For example, some foraminiferans, which are plankton with tiny, coiled shells, shape their shells in one direction in cold water and in the opposite direction in warm water. The percentage of left-to-right coiled fossil forms of these plankton in the ocean's sediments can tell us the sea's temperature at a particular time in the earth's history (see plate 7).

An example of how plankton balance can go wrong occurred in the 1960s. Lake Washington, an urban lake in Seattle, was becoming eutrophic: it was so rich from fertilizers in runoff and sewage that a blue-green alga bloomed in the freshwater plankton. That alga flourished, taking nutrients away from normal plankton, clogging the gills of fish, and reducing the clarity of the lake, so other organisms in the lake suffered. Swimming areas were closed. The lake was well on the way to becoming a muddy, sterile body of water, much like Long Island Sound is today, with few fish and green growing plants. Based on dire predictions of scientists monitoring the situation, the surrounding community was able to stop the processes leading to eutrophication by diverting sewage and runoff. Lake Washington is now a scientific success story: instead of being a turbid, dead lake, it is clean and clear, and the plankton are back to their normal state.

Fifth, plankton can be lethal, even to some of the organisms that eat it. Or the plankton's poison can be collected by and concentrated in the animals that eat it. One classic example is paralytic shellfish poisoning (PSP), or "red tide." PSP happens because filter-feeding shellfish we eat, like mussels, clams, and oysters, collect the poisonous plankton. Carnivorous octopuses don't accumulate these poisons and so aren't poisonous to humans. The organisms that cause PSP are one of those pesky groups that seem to be both animal and plant, and they bloom in such huge numbers, they cause the water to be a rust-red color. In 1793, while exploring the North American West Coast, the crew of explorer Captain George Vancouver ate some mussels from the shores of British Columbia. Four of the crew became sick, suffering numbness and tingling of the arms and legs followed by paralysis, and one died. In modern times, few people are affected by PSP, since commercial shellfish harvests come from safe water and at safe times, and the public is notified of red tide blooms. Red tides usually occur in the summer: there's an old, usually reliable adage that says to only eat oysters in months with an R in them. Red tide can kill marine animals too; a red tide bloom off the coast of Florida recently eliminated a local population of pygmy octopuses that ate infected shellfish.

While dead plankton and its waste products are constantly falling down through the ocean to the bottom, continually adding to the sediment, living plankton must remain in the rich, well-lighted surface layers. This is particularly important for a plant or an animal with chlorophyll, which needs the energy of sunlight to drive photosynthesis. Light is absorbed in the upper 1000 feet (300 m) of the ocean, and the lower depths are essentially black.

Although plants and animals of the plankton, including cephalopods, use several methods to stay near the surface, many members of the plankton have no special adaptations for staying afloat. They slowly sink down through the ocean's surface layers and die when they get too deep to photosynthesize or find food. But they are fast and prolific spawners, and they reproduce before they sink too deep. Their eggs and sperm are buoyant, so they float to the surface, where fertilization takes place and the whole life cycle begins anew. Organisms that go through this cycle (such as diatoms) are usually small.

Other planktonic organisms are adapted in some way to making themselves buoyant or to hover, maintaining their position. These adaptations can either be active or passive. One common passive method is to increase the ratio of the area of the outer surface to the volume, thus making the animal or plant more buoyant or likely to sink more slowly: for example, some organisms have spines or other projections on the outer surface that provide resistance to falling through the water. Octopus paralarvae have bundles of bristles, Kölliker's organs, sticking out of the skin that may do this. And, paralarval Atlantic long-arm octopuses (*Octopus defilippi*), which have a specialized, long, third arm pair, can extend these arms and drift in the surface waters. Planktonic octopus paralarvae with these adaptations look a bit different from the adults, but they are still recognizable as octopuses.

Another way members of the plankton increase buoyancy is to produce a substance that is lighter than water and retain it in their tissues. Many animals and plants of the plankton do this by secreting a type of oil, and oil is lighter than water. This oil eventually winds up as the hydrocarbons of petroleum deposits. The oil-rich blubber of whales helps offset the sinking of the whales' heavy bodies and also keeps them insulated from the cold. Many planktonic organisms have oil globules in their cells, which are visible when they are examined under a microscope.

Another substance used to keep plankton buoyant is ammonia, again lighter than water. Ammonia is primarily used by the large squid species, including the giant squid (*Architeuthis dux*), in their tissues, although the glass squid (*Cranchia scabra*) concentrates ammonia inside a special organ. The ammonia in the tissues of these squid makes the living or dead animals smell pungent. Dead or dying giant squid floating on the ocean's surface smell particularly foul. The ammonia in these giant squid also makes them inedible—there will be no giant squid calamari. Clyde Roper (1984), expert on the giant squid, once sampled a cooked piece of this monster and remarked that it was really terrible tasting stuff.

Other sea creatures use jelly and air as flotation substances. The mesoglea, or jelly layer, of jellyfish helps them maintain their buoyancy. We have been recently finding many new deep-water species of jellyfish, which, even though they are light, have to swim to keep from sinking to the bottom. Some jellyfish use air for flotation, such as the Portuguese men-of-war, as do some cephalopods (such as cuttlefish and nautiluses), fish, and marine mammals. Cuttlefish have an internal shell of porous air-retaining calcium, a cuttlebone, and nautiluses have a coiled shell with air-filled chambers.

Another trick to increase buoyancy is to decrease the density of one's tissues. The few mid-water cephalopods, those that live just above the deep bottom, are usually gelatinous, with tissues close to the density of the deep water they live in. The flapjack devilfish (*Opisthoteuthis californiana*) and other octopuses like it are sometimes called jellyfish octopuses, because their flabby, gelatinous bodies remind researchers of jellyfish.

Many of the planktonic animals, including young octopuses, swim to maintain their level in the ocean. The methods of swimming are as diverse as the planktonic organisms themselves. Some general methods include beating of the hairlike cilia, thrashing with one or several whiplike flagella, paddling with fins or feet, pulsing, oscillation or undulations of the whole body, or the jet propulsion used by paralarval octopuses. Hatchling octopuses, like squid, may use jet propulsion to raise them in the water column, and then take a long glide to rest before pumping again.

It is hard for us to be able to imagine a baby octopus living in the plankton, constantly avoiding sinking into the abyss, watching for gigantic predators that would scoop it up, and deciding which way to swim. We humans are used to living in two dimensions, not looking overhead much, looking forward from our head, and pivoting to see behind us. Our eyes are binocular—both eyes are directed forward with the fields overlapping. Many predators have this arrangement, because it is the best method for tracking mobile prey by sight. But living in the plankton is a three-dimensional experience. Animals in the plankton in open water, such as octopus paralarvae, need to look up and down, left and right, forward and backward to avoid getting eaten by hungry predators. So these animals have big, wide-angle eyes that are adapted to seeing in all directions.

For the most part, living in the plankton means living in a miniature world. Most planktonic organisms are tiny: a hatchling common octopus is just over 0.1 in. (3 mm) long. Within a drop of ocean water, there is a tiny world, with plants, grazers, and predators. Young octopuses have good vision and some swimming ability, so they can jet after tiny crustacean larvae and capture them with their stubby arms. They also try to avoid getting eaten by sinking quickly when a predator looms. Many do get eaten, but the species usually is not extinguished because there are so many hatchlings. Many of the predators are tiny themselves.

Since living in plankton means being carried by the ocean's currents, paralarval octopuses go wherever the currents carry them, and such dispersal is good because it can lead to a wide species range. Yet, paralarvae from octopuses in shallow waters carried by currents out into the mid ocean will probably die. The mid-oceanic islands of Hawaii and Bermuda provide interesting parallels and contrasts for planktonic dispersal of octopuses. Bermuda sits in the mid Atlantic, but it is swept by an arm of the Gulf Stream, the largest river on earth. Although Bermuda is at the same latitude as South Carolina, it is kept subtropical by warm currents that sweep by equatorial South America and through the Caribbean before hitting it. After flowing past Bermuda, the Gulf Stream flows to the cold North Atlantic.

Bermuda is colonized by planktonic paralarvae and juveniles carried to it by the Gulf Stream, including those of common octopuses. So the octopuses there are the same species as in the Caribbean, though fewer. Also, the planktonic offspring from marine animals living in Bermuda are carried northward, often dying in the colder waters or unable to find a place to settle. Some live because they are caught in local currents and stay in the warm bays of the islands themselves. They have to land on an island or a continent to live.

Hawaii, on the other hand, sits in the mid Pacific, isolated in the center of a mid-oceanic gyre of ocean currents that circle these islands. As a consequence, plankton including young octopuses tend to stay in the area, and currents from other islands rarely reach Hawaii. According to the theories of biogeography—essentially why an animal lives at which places—Hawaii should then have a high percentage of endemic animals (those only found there), including octopuses, and it does. Because of the currents circling the islands, planktonic offspring from several octopus species remain there rather than getting carried elsewhere or lost.

Living in the plankton exposes the octopus paralarvae to a great chance of getting eaten because there's no place to hide, and they do get eaten. Out of the 50,000 eggs laid by a common octopus, it only takes one male and one female hatchling to survive in order to reproduce and maintain the species. And, in fact, if more survived, the ocean would be awash with octopuses.

Predators in the plankton are varied in their methods of catching and eating prey. Jellyfish trail out their long tentacles that are studded with lethal stinging cells. Schools of millions of fish, like herring, snap up billions of planktonic animals each day. Large whales consume plankton like a vacuum cleaner sucks up dust. To get an idea of how menacing these planktivores are, picture a paralarval octopus, just 1/4 in. (0.6 cm) long, as a potential meal for a typical jellyfish, 3 ft. (1 m) across with tentacles up to 60 ft. (18 m) long. Such large animals as jellyfish, whales, or whale sharks eat tremendous numbers of tiny planktonic animals just to survive, including the tiny paralarval octopus.

But the large planktivores are not the main predators of plankton animals. That distinction belongs to arrow worms, which are not worms at all but are members of their own phylum, Chaetognatha. The second most abundant animal group in the sea, they are about 1 in. (2.5 cm) long and have paired fins and a tail for swimming. They have eyes and cilia for detecting their prey, mostly tiny crustaceans called copepods, which they grasp with spiny mouthparts resembling the jaws of the alien of movie fame. To an octopus paralarva, an arrow worm or a ctenophore would appear to be the size of a great white shark (see plate 8). We actually know little about how octopus paralarvae live in the tough world of the plankton and almost nothing of their ecology there. (For insights, you may wish to read Roger Villanueva and Mark Norman's substantial 2008 review.) What little we know comes from a few precious observations in the wild, as well as from field collections obtained with a plankton net and from raising a few species in the lab.

Scientists collect plankton, including paralarval octopuses, by pulling a fine-meshed net through the water. There have been many modifications to the old-fashioned hoop-type plankton net utilized by naturalist Charles Darwin on the *Beagle* in the 1830s. Devices on nets can now record the depth they collect from, characteristics of the water such as temperature and oxygen, and even the volume of water strained by the net, allowing calculation of the original concentration of the plankton collected. Unfortunately, as some scientists have learned in their hunt for planktonic giant squid, nets towed through the water are impossibly harsh on planktonic creatures, especially tiny gelatinous octopus paralarvae. Most paralarvae die from the rough contact with the net, the change in pressure or temperature as they are brought up, being mashed together with thousands of other organisms at the bottom of the net or even just from being lifted out of the water. When placed in a shipboard aquarium, some may live for a few minutes, giving us a tantalizing glimpse of their natural behavior. Paralarval octopuses have been attracted to lights hung overboard from a boat at night and then dipped with special nets, which is the most effective, least harmful method of collecting these tiny fragile creatures.

Such collections of preserved paralarval octopuses, either fresh or in museums, give us clues as to what they look like and what they eat from our microscopic examination of their stomach contents. But most behavioral observations come from rearing a few species in the lab. To date, paralarval examples of only a few species have been reared, including the common octopus, the giant Pacific octopus, and the Caribbean pygmy octopus. They are so tiny and their physiology is so specialized that the task is very difficult. Little Californian two-spot octopuses, which hatch out as benthic young, were used by David Sinn and colleagues (2001) to trace the development of personality, and are a good model of early development of cephalopod behavior.

From examination of many preserved specimens, scientists have outlined the specific adaptations that paralarval octopuses have for living in the plankton. These are possession of statocysts (balance organs, necessary for maintaining position and posture relative to up and down), a "lateral line" along the body that gives them touch sensitivity, large eyes, jet-propelled locomotion, an ink gland, and very fast growth. The few chromatophores they have at this stage are usually contracted, making the paralarval octopus nearly transparent. The large bulbous eyes are used to find prey and see predators. The prehensile arms of a paralarval octopus (see plate 9) are much shorter than those of adult animals, but they are still used to seize prey and are used as rudders and keels when swimming. When octopuses hatch, they have about a day's supply of food in the remains of their nutritious egg sacs, but after absorbing this nutrition, the newly hatched paralarvae need to eat quickly. Like adult octopuses, they already have a working ink gland stored in an ink sac and can poof ink in the faces of planktonic predators. And their growth rate is phenomenal.

For baby octopuses, the time before settling—leaving the plankton and drifting to the bottom of the ocean—must be weighed against the time to disperse. There are advantages and disadvantages to settling early and settling late.

Keeping paralarval octopuses alive in the lab is a difficult task, but it allows scientists to observe the behavior of hatchlings, if even briefly. Among the major challenges to raising newly hatched octopuses in captivity, the first is simulating the open-ocean environment. Paralarval octopuses are not used to swimming into walls, like the vertical sides of aquarium tanks. Scientists at the National Resource Center for Cephalopods, in Galveston, Texas, and at the Seattle Aquarium have used a coating of a tiny checkerboard pattern on the tank sides to help tiny octopuses see the walls and avoid jetting into them. Tanks have also been designed with circular walls and upwelling circular currents to keep the drifting paralarvae from hitting the walls or sinking to the bottom. New or recirculated water must be drained from the tank without sucking in the paralarvae, so tank drains have to be screened thoroughly.

Another problem of raising paralarvae is providing them with adequate food. In the wild, they eat tiny copepods, shrimp and crab larvae, and larval fishes. These organisms must move (be alive) in order to attract the attention of the paralarvae. At the Seattle Aquarium, staff worker Susan Snyder fed her paralarval giant Pacific octopuses tiny bits of shrimp meat but had to keep the pieces moving for the tiny animals to eat them, and she had to be meticulous in keeping uneaten food off the bottom to prevent fouling the tank. Providing live food for paralarvae is a distinct challenge. Villanueva and Norman (2008) fed common octopus paralarvae the young of several crab species. Paralarvae are also susceptible to disease, primarily a bacterial one presenting as white patches on the arms and body, which can be treated by chloramphenicol. Disease may be a reflection of the water quality, especially if you use seawater from an urban harbor.

Another problem in raising paralarvae is their cannibalism. Octopuses have no recognition of their own species, and will eat smaller individuals of their own species (conspecifics), whether parlarvae or adults. Each female produces hundreds of thousands of paralarvae, and in the ocean, dispersion by the surface currents usually keeps them from encountering one another. But in the lab, animals are usually reared together and are relatively crowded, giving them the opportunity to eat each other (see Julio Iglesias-Garcia et al., 2007).

Although few species of paralarval octopuses have been raised in the lab, scientists have made a number of valuable observations on newly hatched paralarvae octopuses of species that may live for up to a week or so before dying, probably from starvation. When one aquarium worker attempted to keep giant Pacific octopus paralarvae alive, he fed them bits of krill, which sat on the surface film of the water a few minutes before they sank. The young octopuses turned upside down and actively searched out the food particles floating on the surface, which is known as neustonic feeding.

When Villanueva gave them crab zooea larvae in 1994, he watched the octopuses go through the positioning, orientation, and jet attack sequence that John Messenger described in 1977 for cuttlefish. He also observed that the paralarvae grasped and probed the food for several seconds before moving it along the arms to the mouth, suggesting that they contacted the food with the chemosensory cells in the suckers to establish whether it was edible before ingesting it. He also found that the hatchlings were selective: they would eat krill more often than brine shrimp or larval fish, and survived much longer when fed that food than when fed the other foods.

Richard Ambrose (1981) observed the behavior of Verrill's two-spot octopus hatchlings both in the field and in the lab. He found that they swam backward except when catching prey, using their water jets as a primary mode of movement much as squid do, and that they kept a near-constant upward tilt of 45 degrees, with arms trailing down. At hatching and afterward, they exhibited a strong positive phototaxis—swimming toward the light—as do most octopus paralarvae. This behavior is likely because the paralarvae need to swim upward to keep from sinking down from the rich surface layers of the ocean. When swimming toward a prey item, such as a copepod, the paralarva used a forward swimming movement in a straight line toward the prey item, which it then grasped with its arms. These paralarvae also ate brine shrimp, but they only lived six days, probably because of a nutritional deficiency.

As they grow in the plankton, the paralarvae quickly change in shape as well as size. We usually describe their growth as a percentage of weight gain per day. In the paralarval stage, octopus growth is about 5 percent per day. To put this in perspective, at this rate a 5-lb. (2.3-kg) human baby would put on 1/4 lb. (0.1 kg) the first day, and at the end of the first month would weigh 22 lb. (10 kg). This very fast growth rate is rarely achieved by any other organism.

The change in form is necessary because the paralarvae are going through a big transition in living style as well—they are soon going to settle to the bottom and begin the life style of an adult octopus. The rest of the body grows faster than the eyes, which therefore shrink in proportion to the body, as in human children. Although still used for sighting prey, predators, and conspecifics, the eyes will no longer have to be able to see underneath the body when the adult octopus is sitting on the substrate, or sea bottom.

The arms grow longer. Common octopus hatchlings have arms 37 percent as long as the mantle, but at settlement, the arms are 91 percent of the mantle length and are still short of the up to 400 percent of mantle length at adulthood. Benthic adult octopuses use their much longer arms in different ways from the paralarvae. The arms are used for probing under rocks, throwing arms in parachutelike webovers in prey capture, for defense, and in mating.

In a quick transformation just before settling, the maturing paralarvae grow more chromatophores and suckers on their arms, compared to hatchlings having only a couple of working chromatophores and a couple of suckers per arm. Adult octopuses live in a much more complex and varied environment than the paralarvae do, and they have to develop skin patterns and papillae for camouflaging, and suckers for manipulating the environment and catching prey. Sigurd von Boletzky (1987b) suggested a further transition of the hatchlings, their brain size. The brachial lobe, the brain area that controls the arms, enlarges, and there may be changes in the memory area of the brain, the vertical lobe, which helps the adult octopus to adapt to changing environments, to capture various prey animals, and to escape from different predators.

Another change from the planktonic to the benthic mode of life takes place in the octopus's cardiopulmonary system. Since planktonic paralarvae swim constantly, using their water jets as a squid does, they must get oxygen out of the water at the same time as they are swimming. Therefore, the water jet is used for locomotion as well as for respiration. Benthic octopus adults are rather poor swimmers: human scuba divers can frequently keep up with or even pass them. Martin Wells and his students (1978) showed one reason for their poor swimming: an adult octopus goes into temporary cardiac arrest and oxygen debt while swimming—the three hearts literally stop beating. After a major swim, the octopus needs to rest a while and "catch its breath." In addition, the pigment in the blood of the octopus is the oxygen-binding hemocyanin, which is less efficient than our hemoglobin. But to date, no one has measured and reported the cardiorespiratory efficiency of paralarval octopuses.

We scientists know little about the important transition from a pelagic to benthic life for octopuses. It's a big ocean out there, the paralarvae are still tiny, and only when someone finds them by accident at the right time do we gain any insight into the process. Maybe as they get bigger and heavier, they sink. Maybe as the brain develops, different areas mature and dictate bottom-seeking behavior rather than planktonic behavior.

The paralarvae that are ready to settle and take up a benthic life have to be changed both physically and mentally from their life in the plankton. Species of octopus paralarvae have been studied in groups. Red octopus paralarvae have been observed from an ROV (remotely operated underwater vehicle) at a depth of 500 ft. (150 m), hovering in mid water off Los Angeles in the Catalina Channel. Since the observed paralarvae were far below the plankton layer, they had probably achieved the right size for settlement and were on their way to the bottom, gradually drifting down en masse. Many adults were seen on the bottom at this location, and thousands of these paralarvae were seen together in a loose group. Based on the results of rearing experiments in the laboratory, the paralarval size may be the most important factor in determining whether an octopus is ready to settle or not. It is possible that these red octopuses were the correct size for settling but that they were in water that was deeper than normal and therefore were taking a long time to get down to the bottom.

Settlement of paralarval octopuses to the bottom is a major transition. They drop through the water to the bottom of the ocean, carried by currents, and arrive in new, strange areas. They have new adaptations in their bodies, to help them cope. While coming to new areas may be good for the species, for an individual it can be hazardous. The paralarvae may drop down into the ocean's depths where it is too deep, dark, and cold for nearshore species to survive. They may drop onto animals, such as sea anemones or corals, that eat them. They may drop into a school of fish that might eat them before they can get to a protective den. They may settle onto a coral reef, a boulder field, a bed of gravel, or a field of mud. Still, widespread species like the common octopus and the day octopus survive and reproduce in these new environments. They can camouflage on mud or sand as well as on diverse coral reefs. And they can find or make a den there.

When octopuses settle, they must be able to catch many types of prey in the new environment. They may have to find crabs, clams, or snails hidden under rocks or buried in the sand. They may need to eat snails washed off rocks from the intertidal zone, or catch crabs that live only on wave-washed intertidal rocks. They must be able to subdue their prey and prepare them for feeding. But it's not quite so all-or-nothing. Jennifer found that the newly hatched Caribbean pygmy octopus swims at the surface and clings to the sides of aquarium tanks for several weeks, and pelagic paralarvae can delay for weeks before settling down for good. The juveniles' body shape, characteristics, innate behaviors, and learning ability are all matched to the place where they will live, up in the surface layers or down on the bottom.

Juvenile octopuses, those that have settled to the bottom or those that hatch out as big enough to live on the bottom right away, probably use several methods to determine whether a habitat is suitable for settling. They may use sight to look for suitable den sites and explore the substrate with their suckers, but settling octopuses are negatively phototactic (avoiding light), seeking dark places. They also must look for prey and watch for predators. The temperature, clarity, or salinity of the water may play a role in their choice, and octopuses can detect these characteristics of seawater. Paralarvae can delay settling to the bottom, but not for long, if they find that they are in a place that is not to their liking.

Aquarium workers such as Snyder raising giant Pacific octopuses found that the paralarvae alternated life styles before settling to the bottom for good, indicating the transition may take some time. The paralarvae still swam as they would in the plankton but gradually spent longer and longer periods attached to the walls of the tank or its bottom over several months. Snyder observed that settlement by one paralarva induced other paralarvae to settle nearby, which appeared to be a social influence on a nonsocial animal.

Martin Wells and Joyce Wells reported in 1970 an interesting settlement by the day octopus in Hawaii. Thirteen juvenile octopuses were found clinging to the underside of a buoy marking the end of a shark-fishing line. They knew these animals were newly landed because the line had been set for a short period. The buoy was set in 210 ft. (63 m) of water, making it highly unlikely that the little octopuses had crawled up from the bottom. Besides, they had a mantle length of just ½ in. (1.3 cm), the same size as the smallest day octopuses found on the bottom. And so we know that these octopuses can live in the plankton until they have reached ½ in. (1.3 cm) mantle length and that they probably settle at about that size. We also know that they sometimes, perhaps often, make mistakes where they choose to settle. Natural selection will probably eliminate those that don't make good decisions.

The behavior of newly settled juvenile octopuses and those of newly hatched large-egged octopuses that don't drift in the plankton is remarkably consistent among the many species of octopuses. Most are nocturnal. They can change color and even produce skin display patterns. Juveniles of species that as adults have fake eyes on their skin, or ocelli, have them at the juvenile stage as well, and all can use the adult combination of walking and swimming. They hide in tiny shell or crevice dens or make their own. They catch prey, preferably crabs, but now adult prey rather than larvae. They have moved on in the life cycle.

3
Making a Living

For all animals, getting and eating food is a critical part of their life. Octopuses are generalists, so it's difficult to study this important behavior category in these animals. They are carnivores, and consume a wide range of animal species. They don't eat kelp or other algae or marine plants since they don't have the enzymes needed to digest them. Pretty much any marine animal on or near the bottom of the ocean is fair game. The prey can be bigger than the octopus, since the octopus's venom is deadly to most animals in the environment.

But octopuses do tend to eat more of specific groups. Researchers from all around the world have noted that octopuses eat a lot of crustaceans. The kind of crustacean varies with the octopus's size and location: pygmy octopuses in the shallow Caribbean Sea eat little hermit crabs, day octopuses of Hawaii catch small coral crabs, and giant Pacific octopuses on the west coast of North America consume Dungeness crabs. Octopuses also eat a lot of mollusks. For a group of common octopuses living near a reef full of mussels in South Africa, Malcolm Smale and Patricia Buchan (1981) found that the mussels were the most common prey. For a giant Pacific octopus on Vancouver Island, digging up clams was its favorite way to find food. For common octopuses in Bermuda, tiny file clams (*Lima* spp.) were a preferred snack.

The first step in an investigation of what octopuses eat is to find out what specific animals are being caught and eaten. There are three common techniques for attempting to get answers. First, in the lab, you can feed octopuses different prey species, and if they eat them, then you've established their choices, right? Not necessarily. If you were locked up and fed turnips and oatmeal, you'd eat them, but as soon as you got out, you'd probably pass them by in the supermarket. Also, you may love fresh figs but don't get to eat them often because they're not available. Octopuses may choose their foods in the same way.

Ambrose verified this contrast between accepting and choosing prey species with his two-spot octopuses in 1984. He compared what they ate by getting a list of their local snail and crab prey in the wild from their middens with what they accepted in the lab. After years of fieldwork, he established that at Bird Rock in the Catalina Islands, they eat mostly top snails (*Tegula funebralis*). But in the lab, they took crabs first. A comparison of choice and availability showed that these octopuses didn't find many preferred crabs and mollusks out in the ocean, and that probably there just weren't many crabs of the right size at Bird Rock.

A second way to find out what octopuses select to eat is to sample their middens. Most octopuses eat the soft parts of the prey and throw the shells or skeletons out in front of their den. The midden-sampling approach works relatively well for hard-shelled prey like clams and crabs but doesn't apply for soft-bodied animals like polychaete worms that have few remains. Also, shells dumped onto middens don't stay put. We watched wrasse fish eat the shell remains of crabs thrown out after a meal by the Hawaiian day octopus, and charted the removal of light pieces of crab shell by waves and currents for the common octopuses. Ambrose (1986) made artificial middens near his two-spot octopuses' dens, and watched hermit crabs take away the snail shells to use as homes. So visiting an octopus den once a week to assess intake by counting hard-part remains offers an incomplete picture of actual meals. The only way to know consumption from sampling leftovers is to know about the relationship between what the octopuses ate and what you found at their home within a couple of days.

Another way to find out what an animal has eaten is to kill it, open the stomach, and sample the contents of the gut; you can also examine the stomachs of already dead museum specimens. But this approach doesn't yield an accurate description of prey choices, because octopuses only eat the soft parts of prey: they scrape flesh away from the shell or skeleton, and they digest some of it before they eat it. What the researcher has to work with is often semidigested lumps of tissue. Biochemical serological analysis is needed to get a clear idea of what the food species is and where there was a different array of species than the midden samples of other octopuses. And you've killed the octopuses to find the answer to a behavioral question, which is a drastic solution, and you have only one sample.

The underlying problem with the question of prey choice is that octopuses eat a lot of different prey species. Ambrose (1984) found that his two-spot octopuses from Bird Rock ate fifty-five prey species over several years.

Octopus Garbage Heaps

I conducted a study (1991a) of what common octopuses ate in Bermuda and what I found outside the dens a couple of days later. A team of volunteers followed two octopuses constantly from dawn to dusk for about ten days. Each day, we recorded where the animals went and what they ate outside their dens when they stopped to snack. We picked up the shell remains, and recorded what they took back to their dens. We then recorded the daily disappearance of prey remains from the den midden caused by sediment sifting over them or waves washing them away. After ten days, I dug through the midden to see what had stayed and what was buried. I remember coming by one den and seeing the eight bright blue shells from a chiton meal (a chiton has eight shell valves), then watching on subsequent days as the shells slipped down the midden slope and gradually got covered by debris.

From these observations, I made a flowchart of food fate, from capture to visible prey remains at the den. About one-third of prey was eaten in places other than the den, but there was no size selection. Both big and small prey were carried home, depending mostly on how far away the octopus was when it found the prey. Size mattered after pieces hit the midden, since small, light crab shell pieces were more likely to be washed away than heavy zebra mussel (*Arca zebra*) shells. Ultimately, I could backtrack from ten shells of six prey species in the midden visits by a researcher every five days to how many of each species were actually eaten. So I know how much food was probably eaten by an octopus that had a given number of prey remains.

—Jennifer A. Mather

We found twenty-eight prey species for twelve common octopuses in one small bay in Bermuda over four weeks, and we have collected seventy-five prey species so far from the common octopuses of Bonaire. In addition, what an octopus of one species takes as prey varies across the range of that octopus. Giant Pacific octopuses, for example, range from California to Alaska in North America and over to Japan. In Japan, they eat fish, shrimp, and crabs; on Vancouver Island, they specialize in crabs, cockles, and Pacific littleneck clams (*Protothaca staminea*); in Alaska, they eat crabs 75 percent of the time. In each of these cases, there were remains of over twenty-five prey species in middens of a small group of octopuses. Clearly, octopuses like variety, and also sample a different variety in different areas.

Foraging theory developed in 1986 by David Stephens and John Krebs sets forth that any animal must maximize its feeding efficiency: it has to get the most energy from its prey and spend the least energy finding, catching, and handling it. How do octopuses manage this tradeoff, and does this tell us why they eat what they do? The fact that they shelter in dens makes them "refuging" predators, moving out from central shelter to hunt and then returning to feed, leaving us the convenient midden heaps. Their energy is spent on search, capture, and handling of prey and then on digestion of food. Since a refuging predator spends a lot of time searching for food (see plate 10), taking whatever likely species it finds, this behavior partially accounts for the wide diet.

It's partially correct that octopuses simply take what's there. We noticed that different midden heaps in Bermuda were full of crab remains or shells of file clams. The number of remains of different prey species of the giant Pacific octopus varied with den location even at the same place, and an in-depth analysis compared abundance of prey species in den remains and around the home. In general, there were more individuals of one prey species available—say, red rock crabs—where their remains showed that the octopuses had been eating more crabs.

But availability isn't the only deciding factor. Octopuses learn well. It's possible that the accumulation of shells in front of the den resulted from the octopus having learned a particular hunting strategy and going to a particular area where it worked well. That possibility may have prevailed for the red octopuses we found in beer bottles. With the chance to live in a mud-bottom habitat because of these man-made shelters, they specialized in eating the abundant olive snails available there. Back on the rocks, they might be generalists in their eating habits, but they haven't been studied there. Learning and habit might override a take-it-if-you-find-it approach.

Learning may account for the fact that the octopuses we studied in Bonaire were what we labeled specializing generalists. We were studying squid, but we couldn't ignore the octopuses moving under us as we squid-watched. So we sampled the remains of what they ate, and over three years there were seventy-five different crab, clam, and snail species. But some of the octopuses specialized, often narrowly. One ate immature queen conch snails (*Strombus gigas*), and another dug the fragile pen shells (*Pinna car-*

nea) out of the sandy mud. Habitat couldn't account for these differences, since they were all living in a mixed rock-rubble-sand-mud habitat. Maybe each octopus learned a special foraging technique that fit particular species. It's ironic, though, that the species is generalist but some individuals are specialists.

Besides learning where to find prey, octopus choices of areas to hunt may be restricted by mammals in the area, like sea otters, both as predators and as competitors for the same food species. We wondered whether a predator would learn where to find common octopuses. The octopuses we watched in Bermuda had very small home ranges but shifted them on average every ten days, maybe because they had used up all the easily accessible food, or perhaps because they were working out a balance of energy spent in hunting and energy gained by eating. We wondered whether they shifted home ranges because ten days is about right for a predator fish to catch on to where an octopus lived. And in fact, this speculation has foundation. As we got to know the area as the local fish might, we learned to gather clues to an octopus's possible den site. A fish might find an octopus in this way, then grab it when it's out hunting. The octopuses didn't spend much of their time carefully balancing energy tradeoff and deciding whether this hermit crab or that snail was worth a try. Besides, they were gaining weight. But while hunting, their appearance was disguised and they were always looking for predators. One false move and they would be prey themselves.

Part of the wide choice of prey species may be because of octopus energetics. Unlike homeotherms, such as mammals that keep their bodies heated and spend a lot of energy doing so, octopuses are poikilotherms: their metabolism slows down in cooler water, and they become less active and waste less energy. A major amount of an octopus's energy can be spent on digestion, so energy expenditure for foraging may loom less large in their energy equation. They don't have to eat regularly just to keep going. In the lab, our healthy young pygmy octopuses didn't eat anything at all some days. Octopuses are so efficient at converting food calories to body weight that energy output may be a smaller part of their daily budget than it is for us.

Every researcher has found prey species that their octopuses wouldn't eat. Two-spot octopuses from California and the day octopus of Hawaii wouldn't touch one species of top snail; giant Pacific octopuses didn't eat hairy crabs; and Bermuda common octopuses seldom touched chitons. Perhaps the answer to this issue is preference; we don't know.

Eat or be Eaten

If predation limits octopus foraging time and food intake, maybe keeping an octopus with predators would cut down on both its time out hunting and the weight of crabs it consumed. In Hawaii, I tried to test for this predator influence. I kept two Hawaiian day octopuses at a time in a small outdoor saltwater pond and gave them lots of crabs as prey. As ten days passed, the crab supply got depleted and the crabs got more wary. The octopuses hunted longer and caught fewer of them. When I put a moray eel predator in the pond with them, the octopuses did not limit their foraging time as they should have if they were wary of the predator. Maybe the pond was too small for them. Maybe it was because octopuses don't stay in one territory in the ocean, and if they have no place loyalty, they could just leave if a predator menaces. And maybe it's because moray eels hunt by sneaking in and around the rocks, so not going out wasn't going to help the octopus. This is a good example of the researcher's lament: find the answer to one question and it raises two or three more.

—Jennifer A. Mather

Prey species have evolved many ways to avoid being caught by octopuses. Many mollusks count on their protective shell to save them from predators. While providing some resistance, the shell doesn't stop predation by a variety of animals. Scallops can swim away by clapping the valves together, a successful method for avoiding capture by slow sea star predators but not for avoiding the jet-propelled grab of an octopus. The fragile-shelled, scalloplike file clam hides in crevices; scavenging wrasse can't get them but they are vulnerable to the common octopuses with its flexible arms. Octopuses normally hunt by feeling around in the landscape, with touch and chemical receptors in their suckers probably helping them recognize sources of prey. The approach is an effective one. We timed common octopuses in Bermuda, and it took five minutes for them to snake an arm into a crevice, capture the prey, throw away the shell, and start to digest the meat.

Crabs, being mobile, are a tougher challenge for octopuses to catch than the slower snails and often immobile clams. If you lift a rock, crabs will scrabble out from under it and hide under one nearby, and when you pick up the second rock, they will scurry back under the first rock. Foraging common octopuses and Hawaiian day octopuses are often trailed by wrasse and other fishes as they move across the bottom, and perhaps the fish are planning to eat the small animals escaping from the octopus's menace. We watched a blenny fish in Bermuda go around to the far side of a rock intent on doing just that, waiting for escaping crabs as a foraging octopus slid under the rock. And crabs also pinch. Octopuses—and people—trying to collect them eventually learn to grab the crab from behind. Swimming crabs, like the blue crab, besides hiding under rocks or running away, can also swim away. In response to an escape attempt, an octopus can go from crawling to jet propulsion and can launch a mid-water grab.

The champion octopus evasion technique belongs to some hermit crabs that have a symbiotic relationship with sea anemones. Hermit crabs pull these flowerlike but stinging animals off their perch on the rocks and place them on top of the borrowed snail shells they use as homes. Predators such as octopuses and crabs, when reaching for the hermit crab, are stopped short by the sting of the sea anemone. The anemone benefits from this arrangement: it gets a free ride on this new home as well as castoff scraps from the crab's scavenging. This cooperative relationship has been noted in the lab. A group of anemone-carrying hermit crabs had been kept in a tank at Banyuls, southern France, for months and had gradually dumped their anemones off their shells onto the rocks. In their 1979 study, Donald Ross and Sigurd von Boletzky got some water from the octopus's tank and added it to the crabs' aquarium's saltwater intake pipe. Chemicals in the water from the octopuses must have been carried to the hermit crabs and served as a warning that trouble was around. The hermit crabs rushed over to the discarded anemones, manipulated them off the rocks, plopped them onto their shells, and again were protected.

What we call prey "handling time" varies widely with prey species and ought to but doesn't always influence prey choice. Handling time is the time and effort involved in actually getting at the food within the prey. We come back to the example of crabs: octopuses often choose to eat crabs. Most shallow-water species take crabs in the wild, and the Hawaiian day octopus eats crabs almost exclusively. Octopuses are well equipped to handle most prey. But for many prey species, before it can eat, the octopus sometimes is faced with the "packaging problem." The octopus's strong arms are very good at pulling apart shells as well as pulling animals from hiding or even off the rocks. And in addition to their parrotlike beak that can grasp and bite with efficiency, octopuses also have a ribbon of teeth, or radula, for small jobs. They also produce venom from the posterior salivary gland that can paralyze prey and start digestion.

When the octopus catches a crab or lobster, this arsenal of prey handling tools is supremely effective. A bite or drill through the joint or skeleton and the injection of venom from the salivary gland result in a quick death for prey. The pygmy octopus can subdue a crab its own size very quickly. Marion Nixon and Peter Dilly (1977) studied the remains of crabs that they had fed their common octopuses in the lab. The remains were disarticulated—all the units of the shell were separated, probably by digestion of the tendons, and the meat was totally cleaned out of the individual units. Once in a while, we'll see juvenile common octopuses save a particular part of a large crab until the next day, probably because they already had a full meal. Sometimes they will drill a neat hole into a crab claw, maybe to get better access to the muscle inside the exoskeleton. This approach might loosen the hold of the muscle from the shell plate that controls the thumb of the claw, for easier access.

For octopuses preying on clams, which have solid defenses, the situation is quite different. Because for millennia many different kinds of animals have used penetration techniques to get into mollusk shells, mollusks have evolved bigger, thicker shells for protection. Sea stars pull clams apart and slide their stomach in to start digestion. Crows pick up clams in their beaks, fly up, and drop them onto the rocks to crack the shells. Oystercatchers stab into them with their bill, crabs break them with strong crusher claws, and some oyster drill snails drill holes into the clam's shells. Octopuses are different only in that they have several different penetration techniques: an octopus can pull a bivalve shell apart. But if the clam is too strong, it can drill a hole and inject venom or chip the edge off a valve and do the same, weakening the clam's adductor muscles so it can't stay tightly shut. Octopuses try the easiest way first, pulling the clam apart.

Just as energetics is a good explanation for trying to pin down prey choice, we can turn to energetics to explain why the octopus uses different penetration techniques to obtain food. We found that the common octopus trying to pull apart the valves of a clam used 1.3 times as much energy as it would have spent in drilling. But drilling took much longer: Michael Steer and Jayson Semmens found in 2003 that it took sixty to eighty minutes for a red-spot night octopus (*Octopus dierythraeus*) to drill a clam shell, compared to ten to twenty minutes for pulling valves apart. So we would expect the octopus to try pulling first, and go to drilling if the clam's muscles are too strong. We may also expect the octopus to select clams based on size. Using the pulling technique gives a good energy return, yet the food from a small clam might not be worth the work of finding, holding, and pulling. Generally, small octopuses open and eat small clams. Larger octopuses, on average, eat clams of a size that can be opened by pulling, but the variation in size selection is wide. And one octopus given two clams of the same size will sometimes pull and sometimes drill. So energetics is only part of the reason that octopuses use different techniques in getting at their food.

An octopus drilling through a mollusk shell is a neat trick. Early researchers took it for granted that the octopus drilled by using its radula, because other mollusks, like oyster drill snails, use rasping with the radula to drill into bivalves. But Nixon and Dilly found that hole boring alternated rasping with extension of the salivary papilla, which secreted acid into the shell hole from a different gland than the venom-producing one and dissolved the calcium carbonate of the shell. The octopus also has the challenge of where to drill the hole. Shells are different thickness in different areas, and they contain water as well as a mollusk body. The challenge seems to be guided partly by preprogramming and partly by learning. Common octopuses drill at the shell's edge in clams, for snails they drill high in the upper spiral near the retractor muscle attachment (see plate 11), and for mussels they drill in the center over the heart. Red octopuses and *Octopus mimus* drill over the adductor muscles that hold the valves of the clam together, and giant Pacific octopuses drill into the umbo, the thinnest part of the shell, and over the clam's body. The red-spot night octopus drills at the edges of the shells where they are thinnest.

To study how octopuses choose where to drill, in 1969 Jerome Wodinsky watched common octopuses drilling conch snails right at the location in the spire where the retractor muscle meets the shell and helps the snail pull back into hiding. He covered this area to see whether the octopus would change drilling locations. When he put on a rubber covering, the octopus just pulled it off, and when he applied a dental plastic covering, the octopus drilled through both layers. When he put a metal coating over the spire, the octopus drilled just at the edge of the metal: it knew where it wanted to drill and got as close as it could. He found that female octopuses that were tending their eggs and had no digestive gland activity occasionally ate conch snails. But without salivary gland venom to inject, they didn't drill and so just pulled the snail out.

We found that the giant Pacific octopus uses another trick to get at its food. Pulling apart mussels and Manila clams (*Venerupis philippinarium*) is easy. But Pacific littleneck clams (*Protothaca staminea*) have a thicker shell and stronger muscles to keep their valves together. Roland tested clam strength on a "clam rack." He glued one valve to a plate at one end and a second one to a hook attached to a strain gauge and a screw mechanism. He put more and more pressure, measured by the gauge, into pulling the valves apart, and stopped when the clam just gaped, giving a good strength reading. Pacific littleneck clams were indeed stronger than mussels or Manila clams, and presumably because of this fact the octopuses likely tried pulling and failed, then drilled, though they seldom drilled either of the other species. But they also chipped with their beak at the edge of the valve, again not breaking much shell off but enough to make a hole that presumably would let in venom. And these chips, like by the drilling of the red-spot night octopus, were at the anterior and posterior shell edges as near to the adductor muscles as the octopus could get.

We also tested whether the clam's resistance to the pull would affect what the octopus chose for food. Steer and Semmens had looked at octopus size selection of clams within a single species and had found, as we did, that large octopuses took a wide variety of clam sizes. Our octopuses also ate more mussels and Manila clams than they did the strong Pacific littleneck clams. We wondered what would happen if we took away the problem posed by the shell and opened the shells for them. When we delivered these clams on the half shell to the octopuses, they changed their choices, eating many more of the Pacific littleneck clams and nearly none of the mussels, leading us to suspect that mussels don't taste that good to octopuses but they are easier to open. And Pacific littleneck clams aren't necessarily worth all the trouble, but they certainly taste good if someone else does the work.

Once the octopus eats its food, it continues the process of digestion. Some digestion happens as saliva eats away at the muscles before the meat is taken in. Food that's already begun to be broken down is passed into the gut, where digestive enzymes break it down even more. Over a lifetime, the octopus eats and absorbs more and more food, first to build muscles and later to build eggs or sperm. Since the octopus can't digest fats, they are passed out of the intestine in a string of feces and blown forcefully out the funnel. These fatty substances are snapped up by scavengers like wrasse, because fish can digest and use them. Octopuses have a high conversion efficiency of food to energy, meaning that they are good at laying down body tissue.

We suspect that when an octopus experiences a food shortage, it may just move on, catch all the easily available prey in one range, and then try another range. If true, this would help explain why octopuses don't hold and defend territories. Behavior ecology theory says that an animal only defends an area that has resources it will keep on using. If an octopus samples an area for easily caught prey and then moves on, that area would not be worth defending.

If food gets really scarce, the common octopus may have a very different strategy for coping. It will stop foraging, and if it's a female, it may even develop and lay eggs earlier than normal. This may be because of this species' cold-blooded metabolism and short lifespan. An octopus uses energy mainly for eating and digestion. If the available food isn't enough to keep it going, it might as well get on with its one-time reproduction; it's better to have some offspring than none at all. This set of food-gathering strategies works for the octopus.

4

In the Den

We can think of an octopus home much as we think of our houses. The octopus uses its den for rest, safety, feeding, and a place to shelter young—eggs in the case of octopuses. We have known for millennia that octopuses take refuge in dens, and fishermen have often taken advantage of that habit. Egyptians fished for octopuses 4000 years ago using clay pots for artificial dens. They lowered the jars on strings, then drew them up after a period of time to harvest the octopuses resting in them. Hawaiians dangled a cowry shell just outside an octopus in a den and speared the animal when it came out to the lure.

Most shallow-water octopuses take refuge in a den for long periods of each day—and about 70 percent of their time—over most of their lives. The den can be as simple as a burrow in the mud, a hole in or under a rock, an empty shell, or even a beer bottle. Some octopuses inhabit a previously existing den, but they will meticulously clean it, if necessary. Often the octopus will partly reconstruct the den or will elaborately clean out and modify an existing crevice.

Octopuses may have started taking up residence in dens about 65 million years ago, when they lost their shells. Before that, ancient cephalopods were protected by either an exterior coiled shell or an interior structural shell. Their shells protected them from predators, so they didn't need dens.

An important function of an octopus den is to provide a place for egg guarding. Many female octopuses lay their eggs in a den and frequently block up the entrance to the den, remaining there while they are guarding the eggs. They select rocks or shells small enough to keep other octopuses and possible predators out but large enough to allow good water circulation between them. Female octopuses are very flexible in their choice of den location and construction, possibly because of their adaptability. For instance, in the lab, the Caribbean pygmy octopus prefers to lay eggs in hard shells with small entrances, such as snail shells. But on the sand flats of St. Joe Bay in the Florida panhandle, it settles for cockleshells since there's competition with hermit crabs and blenny fish for snail shells.

We used to think that shallow-water octopuses made their dens in rocky areas or on coral reefs—places with hard substrate of some sort. We now know that a number of octopus species live on sand and mud and make their burrows there, such as the mimic octopus (*Thaumoctopus mimicus*) and the wunderpus of Indonesia (see plate 12). They may even consolidate the sand of the burrows' sides with mucus that they produce.

Many shallow-water octopuses choose dens with hard walls: a crevice in a rock wall, under a boulder, a hole in a coral head or in a rock, a clamshell, or snail shell. So they generally live in areas where rocks, coral, or shells are available (see plate 13). In the Puget Sound area of Washington, giant Pacific octopuses live where there are rock outcrops or ledges or large underwater boulders they can hide beneath known as "monument rocks" that were left behind by glaciers. These octopuses' dens are in areas with fast tidal currents, which sweep away sediments that expose the rocks, carry away food and animal wastes, and bring in a good supply of oxygenated water. The presence of oxygenated water also means that there is a good supply of prey to find on foraging expeditions.

It is difficult to say what makes a good octopus den: there are so many octopus species in so many habitats, and they are so flexible in their behaviors. But most octopus dens have a few basic characteristics: dens are relatively small, and they have a maximum volume two to three times that of the octopus. Giant Pacific octopuses, for example, live in dens of different sizes as they mature: small ones live in small dens and large ones live in large dens. Keep in mind that it's difficult to measure an octopus's den: usually we can only make a rough assessment of the volume of a structure because the dens are hidden away. And the octopus must be out of the den in order for us to be able to measure it.

Shallow-water octopuses choose dens small enough so that they can keep in contact with the walls with their suckered arms. The Caribbean pygmy octopus sits sideways in its chosen cockleshell and holds the valves of the shell closed by pulling on each side. Small giant Pacific octopuses sit upright in their adopted shells, holding onto the shell above and below them. Either way, they keep the shell closed with their strong arms. This contact with several hard surfaces is also seen in laboratories and public aquariums. In the lab, octopuses placed in a bare tank will settle into a corner, either an upper corner where they have contact with two walls and the

A Two-Dimensional Octopus

The shape of the den does not seem to be nearly as important as its overall volume, since octopuses are great contortionists. At the Seattle Aquarium a few years ago, a large female giant Pacific octopus on public display chose to live behind a fiberglass backdrop at the rear of the tank. She had squeezed through a 2 inch (5 cm) opening at the bottom of the backdrop and into the 3-in.-wide (8-cm-wide) space behind it. It was fascinating to see this 30-lb. (14-kg) octopus jammed almost flat into such a restricted, 1-yd.-by-2-yd. (1-m-by-2-m) space. She apparently liked it or at least preferred it to being examined out in the light by the paying public. At night, she would come out from behind the backdrop to eat the live crabs left in her tank for food, but she retreated during the day. To get her back out on public display, I had to tear down the tank, pull her out, and seal off the opening with silicone cement to prevent her from going back in. We named this animal Emily Dickinson, after the shy poet. Because of her shyness, we eventually released her back into the wild, since she wasn't suitable for public display.

—Roland C. Anderson

surface of the water, which acts as a third surface, or a lower corner where they are in contact with two walls and the bottom.

In the wild, octopuses make their dens in a variety of places. Dens have been found in rock crevices or holes in rocks, shells, kelp holdfasts, and even concrete blocks used as mooring anchors for small boats. But small octopuses and juveniles of larger species of shallow-water octopuses often live in empty mollusk shells. Many shell collectors have had to evict an octopus before taking home a prize shell found when they were diving, and shell collectors use the octopus's den middens as sources for shells. Jennifer has found that the Caribbean pygmy octopus in northern Florida much prefers to live in snail shells than bivalve shells, but she found that in the sandy bays where there were few snail shells available in the turtle grass beds, they made do with clamshells.

Octopuses seem to prefer a den with a small opening. Because they have no fixed skeleton, they can squeeze through a narrow opening to access a wider den within. We have often seen giant Pacific octopuses inside a narrow vertical crack in a rock wall. Caribbean pygmy octopuses will choose brown beer bottles with the tapered neck as dens, and small juvenile octopuses sit in the neck, keeping in contact with a hard surface with all arms.

In Hawaii, the day octopus may make a den in coral rubble with an upward-facing opening from which it will peer out, or it may close the opening with coral rocks when it wants protection while it takes a nap. The common octopus also will make a home in coral rocks. These dens and those of most other octopus species frequently face outward to deeper water, maybe to keep an eye out for predators or potential prey. We sometimes saw a Hawaiian day octopus in a den do vertical head bob movements, using the change in viewpoint to estimate the distance to a faraway object or animal, and then jet directly to and capture a crab 10 ft. (3 m) or more away. The octopus seemed to have been on watch for such an opportunity.

Sometimes octopuses make their own dens where natural dens are limited. Like the Hawaiian day octopus, they may dig a den into rubble or under a coral rock or gather rocks together. A giant Pacific octopus making a den under a large boulder will construct it by blowing and carrying out sand and gravel from underneath the boulder. We observed this same den-making behavior in a Hawaiian day octopus that we were keeping in an outdoor pond on Coconut Island in Kaneohe Bay, the research station of the University of Hawaii; it built a den under a cement block. Octopuses can blow sand and gravel out of their den with jets of water from their funnel, or they tuck material in the webbing between their arms and carry it to the entrance. We found that the day octopuses in Hawaii and the common octopuses in Bermuda were good housekeepers; they spent time cleaning out their dens each day. When a den began to look messy, we suspected that its occupant had moved on. When an octopus excavates its own den, it can make it the exact size, shape, and volume it prefers. Some octopuses, like the giant Pacific octopus, build several entrances to their dens, while pygmy octopuses' dens have only one entrance.

Compared with the giant Pacific octopus, smaller octopus species spend shorter amounts of time in any one den. Jennifer found that the common octopus in Bermuda stayed in a den an average of ten days, with a range from one day to one month. One octopus switched between two dens daily. The octopus that stayed longest in its den lived under a fishing dock where crabs came from afar to eat fish remains, so the octopus probably

Octopus Mansions

Most giant Pacific octopus dens have at least one back door in addition to the primary entrance, and sometimes there are up to seven total entrances. I saw this trait in the largest octopus I ever found, which I estimated to be over 110 lb. (45 kg) in weight. It lived under a large flat rock more than 10 ft. (3 m) across in Hood Canal, on Puget Sound. I looked under the rock into the octopus's primary entrance behind an exceptionally large pile of excavated sand and shell remains of Dungeness and red rock crabs. I saw an octopus with suckers at least 3 in. (7 cm) across. I swam across the rock to look in the back door and also saw arms there with large suckers. The octopus filled the entire space underneath the rock—a very large octopus indeed, living in this spacious area. This large Hood Canal den is usually occupied by a giant Pacific octopus, and so scuba divers there have recounted the myth that the same animal has lived there for months or years. In fact, nonbrooding giant Pacific octopuses only live in any one den for about one month before moving to another one. Some dens are almost constantly occupied, but by different octopuses.

—Roland C. Anderson

stayed because of the good food supply. The Caribbean reef octopus (*Octopus briareus*) also only spends days in a den. Brief occupancy in a den makes ecological sense, since the octopuses are likely to eat all the crabs, clams, or snails in their small home range if they spend too much time in a single den. They eat up all the easy-to-catch prey, and then they move on.

Octopus dens attract other animals as well. Giant Pacific octopuses share their dens with several species of fish and invertebrates, such as sea stars, crabs, and scavenging snails. Jennifer found that the wrasse dubbed "Slippery Dicks" were den associates for the common octopus of Bermuda. The day octopuses of Hawaii also attracted wrasse to their food remains, and hermit crabs, perhaps to take the discarded shells of snail prey for new homes. The fish may have used octopus dens for protection against predators through the presence of the octopus, even though the octopus could also prey on them. At Cape Flattery, on Washington's northwestern coastal edge, the midden piles of almost all giant Pacific octopuses are covered with small hermit crabs, likely scavenging on leftovers from the octopuses' food. And these crabs are free to forage in close proximity to the large octopuses, because they are too small for them to bother eating.

A midden advertises the presence of an octopus den. Scuba divers have long used den middens to spot giant Pacific octopuses. Likewise a hungry fish or marine mammal may use a midden to detect its potential octopus prey. Dens are also evident by a pile of natural material, usually sand and gravel, excavated from inside the dens, causing the "volcano effect," since it looks like the talus slope of a volcano emerging from the crater of a den. Den middens are a good source of information for scientists about the food that octopuses have eaten. Giant Pacific octopuses make particularly large middens in front of their dens, a treasure trove for studies of prey choice and food handling techniques.

An octopus usually has a strong grasp on its shell home. Some years ago, while studying Kennerley's venus clam (*Humilaria kennerleyi*), we collected some live specimens and took them back to the Seattle Aquarium. While they are normally found buried in shallow gravel and sand, many live clams were lying on the ocean's bottom, presumably having been dug up by sea stars that were then unable to open this strong-muscled, thick-shelled clam. While diving, we judged whether a clam sitting on the gravel was alive by attempting to pull the valves apart. Confident that we had live clams, we transferred them to holding facilities of the aquarium. On examining the clamshells the next day, we were surprised to find that five of twelve were inhabited by octopuses rather than clams—four by red octopuses and one by a small giant Pacific octopus. They had been holding the shells tightly shut with their suckered arms and were almost as strong as the clams themselves.

Den locations may be the result of many different activities, such as reproduction. One or several males may make dens close to a female, as if waiting for her to be ready to mate or looking for the opportunity to mate with her. In the case of a giant Pacific octopus, dens may be as close as 3 ft. (1 m) apart. When we watched Hawaiian day octopuses to assess their feeding in Coconut Island, we saw a male and female establish dens half a yard (half a meter) away from each other.

Some species of octopuses bear planktonic paralarvae that do not have much of a choice where they make a den if water currents sweep them into poor areas—one reason that octopuses are so adaptable. They have to make the best of a bad situation when they settle to the bottom and there is no readymade den. If water currents carry an octopus paralarva out onto a sand flat and it has to settle out of the water column, it may be dangerously exposed. Although it can camouflage itself by matching the color and texture of the sand, it needs to find or make a den in which to rest and sleep, a place of protection where it can relax its camouflage, which takes muscle contraction. We have seen giant Pacific octopus dens that were created by carrying and blowing sand out from under a sunken log resting on the sand bottom or inside hollowed-out sunken pilings on the sea floor under the Seattle Aquarium. Jennifer has seen common octopuses in Bermuda living in pipes and tin cans.

Lurking in Shipwrecks

Octopuses even live in wrecked ships. If we believe what we see in Hollywood films, every sunken shipwreck has a giant octopus lurking in its murky cabins, perhaps even guarding a sunken treasure chest. Although few sunken ships in the North Pacific have treasure chests, many shipwrecks have one or more giant Pacific octopuses living in them. I write a shipwreck column for a Puget Sound–based dive magazine and have dived on many shipwrecks and seen many octopuses living on or under them. These giant Pacific octopuses can get as large as 400 lb. (180 kg). While they do live in wrecks, they don't attack or kill divers.

Once while diving on the wreck of the clipper ship *Warhawk* in Discovery Bay off Puget Sound, I saw the greatest number of giant Pacific octopuses I'd ever seen on one dive. This full-rigged sailing ship caught fire and went down in 1883, and all that remained were the skeletal ribs of the ship's starboard side protruding from the sand bottom and 100-ft.-long (30-m-long) pile of ballast rocks next to the ribs. Eight giant Pacific octopuses were living in the ballast pile, perhaps because there was little else to make a den out of nearby in the bay, only vast expanses of sand and mud. These octopuses made dens where they could, even though they were closer to each other than they would have liked. Instead of just a home, this was an octopus condominium.

—Roland C. Anderson

Octopuses make their dens in many other human-donated items. Common octopuses of the Mediterranean Sea live inside ancient amphorae, sometimes all that is left of sunken Roman galleons on the sea floor. In the northeastern Pacific, the little red octopus may move into garbage on the sea floor. A female red octopus trawled up from the Friday Harbor area, in Washington's San Juan Islands, was living in an old shoe. This was a female with eggs, so she was quickly named Mother Hubbard by the college class that found her.

Roland once found a giant Pacific octopus living in and blocking the active outfall of a sewer pipe in Tacoma, and a retired professional diver reported seeing several giant Pacific octopuses inside the boat locks between Lake Washington and Puget Sound during routine cleanings of the locks. Octopuses are frequently caught in crab traps in the North Pacific and in lobster pots in other parts of the world. They don't intend to use the traps as dens, but on their foraging expeditions they frequently pause in a temporary minimal shelter to quickly eat whatever prey they've caught, rather than take it home to eat it. And so a crab or lobster trap can become a snacking stop, replete with a handy supply of food. Unfortunately for the octopus, the trap is sometimes lifted to the surface before the animal has finished its snack and moved on. The octopus may then be used as bait for future catches.

Our trash can be useful for supporting octopus populations. On a recent scuba dive in Puget Sound, Roland saw eight beer bottles littering the bottom, and each had a small red octopus in it. Since the beer bottles at that dive site had 100 percent occupancy, the animals were utilizing a new resource. There was very little else for them to hide in other than a few large snail shells that were already occupied by hermit crabs, so lack of suitable dens could have stopped them from living there. Our trash may be increasing the range of this octopus to include areas that have no natural den sites. Secure in its beer bottle home, the animal is temporarily safe from predators. Ironically, the species may lately be facing a housing shortage. Environmental societies in coastal areas are cleaning up trash from public beaches, and as part of these cleanups scuba divers are picking up underwater trash, including beer bottles and cans.

The phenomenon of octopuses living in our beer bottle trash has been useful to researchers. Janet Voight did population studies in 1988 on Diguet's pygmy octopus (*Octopus digueti*), which normally live in shells on intertidal sand flats in the northern Gulf of California. Strings of beer bottles were set out at low tide and monitored for a year. The octopuses captured in this manner were not harmed as they are by other methods such as trawling, dredging, or noxious chemicals squirted into their dens. Dark beer bottles with narrow necks, tapering to an opening less than 1 in. (25 mm) wide, were used. About 14 percent of the bottles examined were occupied by the octopuses.

We have found that red octopuses in Puget Sound prefer stubby beer bottles to the long-necked variety and brown bottles rather than clear or green ones. Furthermore, they prefer "aged" bottles that are covered with a growth of barnacles or sea anemones. These characteristics make the containers darker inside, so perhaps the octopus feels safer than in transparent glass.

Dens aren't just places for octopuses to hide. Other animals are attracted to the site of the octopus's feeding activity, and some animals, such as juvenile parrotfish, aren't found in those sites because of habitat disruption by the digging octopus. Jennifer found that octopus dens in Hawaii attracted scavenging wrasse fish waiting for the scraps after the octopus had finished eating. And the dens also attracted hermit crabs, which need an empty snail shell in which to hide their soft abdomen, and what better place to find one than in the midden of a snail-eating octopus? The dug-up area with a diameter of 3 ft. (1 m) around the den will have less growth of algae, so herbivorous fish won't bother to hang around in that area. These observations remind us that octopuses are part of a complex web of underwater life. They are attracted to specific den sites because they need to avoid being eaten by some animals. And then when there, they change the area, and therefore have influence on other species.

5
Getting Around

Octopuses are a remarkable sight when they are out hunting. They seem to flow across the rocks, holding on with some of the eight arms and extending other arms into crevices and under rocks, occasionally lifting off the bottom and gliding about 3 ft. (1 m) or so. Anyone who has taken a long look at an octopus moving this way across the ocean floor or an aquarium tank has likely admired the animal's fluid movement and flexibility.

There are two kinds of octopus movement: one is about control and coordination—how an octopus gets all the arms to do what it wants them to do. The other is locomotion—how the octopus moves itself around in the environment. Regarding locomotion, the octopus looks as if it has a disadvantage because of its molluscan heritage. Snails and clams just don't move fast, and the molluscan foot, unlike the limbs of vertebrates and arthropods, isn't designed to produce speed of movement. Biologist Richard McNeil Alexander, in his 2003 overview of movement styles among animals, points out that each species makes compromises in balancing the demands for speed and maneuverability with the need to limit energy output. These compromises that octopuses make, based in their phylogenetic history, are intriguing aspects of their means of getting around their environment.

One obvious aspect of locomotion is maximum speed, which varies depending on the medium you are moving through. Cephalopods don't have much speed, comparatively. The cheetah, for example, can manage a top speed of 90 ft. (27 m) per second, which is very fast. Still, the cheetah tires very quickly, and so a prey animal that dodges and runs from it may avoid capture. The pronghorn antelope of the North American prairies specializes in sustained speed and can keep a speed of over 50 ft. (15 m) per second for over ten minutes. Air speed can be faster than land speed since air is not dense, and birds manage sustained speeds near 60 ft. (18 m) per second. A diving peregrine falcon can achieve a gravity-assisted top speed of 170 ft. (50 m) per second, the fastest known. Sea animals are at a disadvantage because of the high density of water: dolphins manage 30 ft. (9 m) per second, and pike can accelerate to 12 ft. (4 m) in less than a second. Squid, the fastest cephalopod, can only move 8 ft. (2.4 m) per second, and octopuses 1 ft. (1/3 m) per second. Clearly, cephalopods can't always outswim fish.

Another demand for locomotion is maneuverability or change in direction. This feature is very useful in the short run, for a predator can be dodged if it can't be outrun. Thompson's gazelles and squid do this very well, since both can pivot 180 degrees while moving just their own body length. For the near-shore squid, this ability is crucial, because predators like barracuda specialize in sudden acceleration and short dashes. For the less maneuverable octopuses, hiding and camouflage replace speed.

While maximum speed, sustained speed, and maneuverability clearly matter to getting around, a wider view of the cephalopod locomotion system is important. But over the long term, energy spent on locomotion is also important, especially when it's calculated as a tradeoff for how far it gets the animal. Octopuses and squid move through the water by propelling water jets one after another through their funnel. Since squid can outjet octopuses, their jet propulsion has been studied more and squid often are used as examples when writing about cephalopod issues. Jet propulsion in squid is not an energy-effective way to get through the water, however, especially compared to the sinusoidal, or sideways-flowing, body bending of fish—an important point because cephalopods have evolved competing with fish.

A second issue for all moving animals is resisting the pull of gravity—staying upright in the case of land animals and staying up within the water in the case of many marine animals and a few cephalopods. Some unusual cephalopods have made drastic changes in their body structure to stay buoyant: for example, the air-filled chambered shell of the nautilus, the gas-filled swim bladder of the football octopus, and the ammonia-filled organ in the mantle cavity of the glass octopus (*Vitreledonella richardi*). These cephalopods have sacrificed speed so they can stay up in the water column, as do octopus paralarvae, which need to stay near the surface where the light is and therefore where phytoplankton and zooplankton grow and where most marine animals are. Other animals, like squid and tuna, don't sink in the water because they keep swimming to stay afloat, and that process is effective while energetically costly. Some animals, like mature octopuses, are heavier than water, but since they mostly crawl on the bottom and only lift off some of the time, buoyancy doesn't matter much.

Alexander points out that evolutionary background is significant in locomotion. Octopuses have to work with energetics and a demand for speed and maneuverability, but they are limited by the basic molluscan model. An animal can only improve the efficiency of the model it has to work with, and the basic molluscan model isn't a good one for speed. Minimizing or getting rid of the heavy shell through evolution has increased the possibility of speed and maneuverability, true for both cephalopods and the shell-less nudibranchs. Gastropods that can stay up in the water include the sea angel (*Clione limacina*), which swims with fleshy "wings," the lion nudibranch (*Melibe leonina*), which undulates a laterally flattened body (see plate 14), and the janthina sea snail (*Janthina janthina*), which drifts on the sea surface attached to a bubble raft. The cephalopods have developed a specialized way to keep moving in the water, through contraction of the mantle, which has been freed from its shell to allow for water ejection and therefore jet propulsion.

Many mollusks, like the scallop and the *Lima* clam, have turned their shell into a movement advantage. By clapping their two valves together and forcing the water out between them at the back, scallops and *Lima* clams (see plate 15), jet propel themselves through the water. The elastic hinge opens the shell again, so the scallop can flap through the water for a couple of minutes. While not well-directed movement, the water expulsion lifts the animal off the bottom and sends it a couple of yards, and when it stops, the flat shell acts as hydrofoil and the animal drifts gradually to the bottom. Mollusks such as scallops swim to escape sea star predators; just the touch of the sea star's tube feet or the saponin chemical in them sets mollusks off in escape responses. Scallops only need to get off the bottom and away, because sea stars aren't fast moving either. But this type of movement isn't successful when scallops are pursued by octopuses, which can lift off the bottom and chase them by jet propulsion. Octopus middens commonly have a few scallop shells, bearing witness to this fact.

The jet propulsion of cephalopods is fairly fast in the short term and is well directed. On the same principle as octopuses but more effectively, the well-muscled mantle of squid, which we humans eat as calamari or squid rings, exerts pressure inward with circular muscle contraction, and the flexible funnel aims the water flow to direct the movement. The connective fibers in the mantle store tension during this contraction, and their elastic recoil pulls the mantle back out, drawing in water for another jet.

The energetic efficiency of jet propulsion is low. The jetting system is designed for another purpose, maximum acceleration from rest. Not only does the fast contraction mean a powerful jet, but also the nervous system of the squid is set up to make it faster yet. A few very large nerve cells, or axons, come from the base of the brain to the squid's mantle, branching extensively. Each giant fiber is different in diameter and placed so that all the areas of the mantle get a nerve signal very quickly and at the same time. This system was discovered by zoologist and neurophysiologist J. Z. Young (1971) in the late 1930s, and because of these giant axons, squid became a valuable species for study at the marine research center in Woods Hole, Massachusetts, near Boston. Much of what we know about how nerve cells work is based on how these squid giant axons perform. The principle of a few big neurons setting up fast escape responses isn't used just by the cephalopods, either. Crayfish have a similar set of giant axons leading to the tail to set up sudden escape tail flips. And the octopus uses sudden jet-propelled acceleration when a predator discovers it despite its camouflage: a couple of jets can take it far enough away so that it can send out an ink cloud screen or hide again decked out in new camouflage. One of our colleagues, Ron O'Dor (1998), described true squid as the Ferraris of the cephalopods and the bumbling little sepiolid squid as the Volkswagens. We'd extend the analogy and describe the octopuses as the Mack trucks.

Cephalopods don't only use jet propulsion for getting across the ocean floor. Squid use fins, ranging in size from little posterior flaps that probably help only in steering, to the wide "wings" of Humboldt squid (*Dosidicus gigas*), or the ribbon of cuttlefish fins that runs along the edge of the mantle. These fins have the same combination of muscle types and elastic connective tissue and use the same kind of motion that fish have. Watching a Caribbean reef squid holding stationary or moving slowly, you can see the ripple of fin undulation and the slow puff of minimal water jets from the down-turned funnel. It's a neat slow-fast system: the fins tuck in along the mantle when the squid jets away (see plate 16). Fishes have a dual muscle system for accomplishing slow-fast movement, a combination of well-oxygenated red muscles that can keep contracting for long periods of time and white anaerobic muscles that need recovery periods. Squid, too, have some muscles used for slow cruising and others used only for escape.

A dual slow-fast movement system is particularly important for octopuses. Normally they move slowly along the bottom by both pulling with the arms and jetting through the funnel. Their speed of progress is one-tenth that of jetting, 10 seconds per ft. (30 seconds per m). Octopuses move slowly, which is logical since they are feeling around for food. Octopuses also use jet locomotion just to get someplace directly. Individuals of some octopus species, such as the giant Pacific octopus, make fairly long migrations, possibly for reproduction (see plate 17). Whenever the octopuses we watched in Bermuda wanted to go somewhere directly, they would lift off the bottom and go several yards with a few jets. In this type of movement, the octopus's flexible body shape is very useful. The giant Pacific octopus, in particular, can spread its arms and flatten them dorsoventrally, presenting a hydrofoil-type surface, which makes motion through the water easier and slows sinking. This motion may look inefficient if you value speed, but octopuses get around effectively.

Regarding the octopus's movement of separate parts such as limbs, we must first think about support systems. No motor system can move an animal without some sort of support to move against or it would simply collapse in a heap. Vertebrates and arthropods have solved the support problem so important in air by evolving skeletons—internal for fish and birds and external for ants and crabs. The limbs are organized as a set of levers with skeletons as the shafts. The benefit of a lever system is that it can get excellent mechanical advantage that gives strength, depending on where the muscles are attached to the bone. The disadvantage is that such a limb can only move at the joints and only in the direction the joint allows. We can't bend the human knee joint sideways, nor move an arm with the flexible grace of an octopus arm. Crab movement is even more limited: most crabs are like medieval knights in armor, clumping across the ocean bottom. The exoskeleton, however, gives crabs protection as well as support.

Without a permanent rigid skeleton, cephalopods have solved the support problem in their own way. William Kier and Andrew Smith (1990) point out that the principle that allows many mollusks to move is that of the muscular hydrostat. A limb or body using such a system stays at a constant volume, but a change in one dimension results in a reactive change in another. It's easier to get the idea of constant volume with variable shape if you envision a water-filled but half-expanded balloon. Press it on one side and it bulges out on the other; squeeze it so it's narrower and it gets longer. This flexibility is used in a lot of molluscan movement, and is the secret behind clams being able to dig into sand. When a clam starts to push its foot down into the beach, the shell is kept spread apart as an anchor while the narrow foot is pushed down into the sand. Then a valve opens and blood rushes into the foot, which swells like a balloon and anchors the animal. The shell valves then close and the now-anchored foot pulls the shell down toward it—a neat trick indeed. Several of these moves, and razor clams disappear into the sand in seconds.

There are no cavities for the blood flow of cephalopods to use to change shape, since they have enclosed arteries and veins like humans. Their muscular hydrostat arrangement follows the same principle of holding volume constant, but depends on the contraction of some muscles acting as a skeleton to oppose others that change the limb's shape or position. Octopus arms have circular muscles around the outside, radial ones from the center to the periphery, and transverse ones spiraling around the length in left-hand or right-hand coils. That's a lot of muscle in the arms, more than half the body's volume. It also means a lot of strength. An octopus can resist a pull of 100 times its body weight. Octopus arms are tubes of a constant volume: radial muscles contract and longitudinal ones extend, letting the animal reach out to grab a hermit crab. And the opposite actions will contract the arm and bring the crab closer to the octopus. Bending resulting from contraction of ventral longitudinal muscles in the arm gets the crab up to the octopus's mouth.

Muscular hydrostats can elongate, bend, or twist by stiffening, relaxing, or contracting different sets of these muscles. We have calculated that each octopus arm is made up of three units—arm tube, sucker stalk, and sucker—all of which can do much the same actions and can do so together or separately. An octopus can extend an arm straight out into the water, reaching toward but not touching potential food and thus not using suckers or stalk (see plate 18). From video analysis of these actions, scientists have deciphered the dynamics of simple octopus arm extension. An octopus starts straightening the arm at the base and unrolls it outward until it gets it where it wants it. Suckers can also be a separate moving unit. Once when we watched a Hawaiian day octopus resting in its den, a tiny foraging hermit crab fell on it. Without moving its coiled arm, the octopus picked the crab up with a single sucker, extended the sucker stalk, and dropped the offender a bit farther away. In a 1994 study, Satoko Seino and colleagues poked an endoscope viewer on the end of a long stick inside the pouch of arms made when a female giant Pacific octopus was laying eggs. They filmed the octopus weaving the long stalks of the eggs around each other before attaching them to the roof of her den, using only movement of the sucker stalks.

The suckers are a particularly intriguing arm unit. William Kier and Kathleen Smith worked out the biomechanics in 1985. There's a cavity, the acetabulum, at the end of the sucker. The animal attaches to some object, such as a clamshell it wants to open, with the sucker rim. Then it decreases the pressure inside the acetabulum, making a vacuum in the sucker that keeps it attached to the object. Because there's local control by a cluster of nerve cells in each sucker ganglion for muscles of the sucker rim, the sucker edge can bend to fit the contours of whatever it's grasping without losing its grip. A fixed sucker, with a rigid edge, can't hold the knobby surface of a crab leg, but the flexible one has no problem holding on to a relatively flat surface. The sucker surface also has touch and chemical receptors, making that surface perfect for exploring and gaining information about whether something is edible. Flies "taste" with receptors on their feet, and octopuses "taste" with receptors in their suction cups.

The octopus sucker can also be folded in half to form a grasping surface. We humans are proud of our thumb-and-finger pincer grasp that lets us hold spoons, pens, hammers, paper—the objects that build our civilization. Octopuses have hundreds of suckers that can make pincer grasps all down the ventral side of the arms. These variable-sized grips are very effective, and octopuses can perform very fine actions with them, like untying knots in surgical silk. Wodinsky (1977) discovered this ability after he'd done surgical operations on several common octopuses, watched them come out from the anesthetic, and then he went home for the night. When he returned in the morning, the octopuses were holding on to the side of their tank with the wound gaping, and the lengths of surgical silk were lying on the bottom, untied.

While the design of octopus suckers is fairly similar across many species, there are some fascinating differences among cephalopods. Octopus suckers, with the flexible rim, are better adapted to holding on to hard and smooth surfaces such as clamshells and crab legs. Squid suckers, especially on the tentacles, may have a row of sharp hooks, or teeth, extending inward from the rim that help them hold soft-fleshed prey such as fish. The deep-sea group represented by the vampire squid has cup-shaped suckers. Maybe in the deep, there is more to explore and less to hold tight. There are more sensory receptors in the rim of the octopus suckers on the outer, distal part of the arms (at the arm ends), with which the octopuses feel, and fewer on the inner (proximal), walking part of the arms.

While an octopus arm has a wide range of freedom of movement, it likely has a smaller number of moves that it can actually make. Still, some of the single-arm moves it makes are fascinating combinations. Repelling a scavenging fish at its den entrance, a common octopus straightens a bent arm suddenly, unrolling it in what we call a "slap." The common octopus and the Hawaiian day octopus can bend an arm back over their own body surface and even into the mantle cavity (see plate 19) to groom itself, running its suckers over the skin surface, lightly holding to and picking up loose skin, dirt, and parasites even from inside its mantle cavity. Few external parasites of octopuses exist, only internal ones like dicyemid mesozoans in the kidney are common, and it's no wonder—all the external ones are groomed away. Among our favorite single-arm movements is the one we call the conveyor belt: with an arm extended out and held steady, an octopus can pass a small prey item such as a hermit crab in a shell from sucker to sucker along the arm's ventral surface to its mouth. Or it can reject a shell that turns out to be empty, passing it out again the same way.

Foraging octopuses have a great deal of independence in arm movement. Moving across the rocky bottom, an octopus will extend a couple of arms in each direction along a crevice. Going to a patch of algae, it will feel along the algae stalks with several arms, grasping with a few suckers if receptors suggest there's prey present. Often the front arms explore and the back ones walk, depending on what direction an octopus is heading. Octopuses can spread all the arms and extend the web down between them, forming an umbrella shape, or a "webover." The webover forms an effective and flexibly shaped pouch. Small prey can be held in a kind of enclosure below the mouth, made up of arm bases and the extended web, with a few suckers of each arm holding on. As the octopus lumbers along the landscape feeling for more prey, even eight arms may be too few, and it may lose one or two of the crabs it's captured. A small crab can dart out and be recaptured by one arm, while the disruption allows two more to escape from the pouch. If a swimming crab lifts off the bottom (see plate 20), a quick Hawaiian day octopus can switch from feeling around on the rocks for the crab to enclosing it by a webover in the water and grabbing it with the suckers.

Coordinating the movement of many arms can be problematic. After all, these limbs must stay disentangled from one another. And not surpris-

Tiptoe

An especially complex and intriguing multiple-arm octopus action is the one I named the "tiptoe." I observed that an octopus wanting to sneak up on prey over a reasonably level surface could extend arm bases down to the bottom, leave the outer half of the arms trailing, and move by holding with and swinging about a dozen suckers per arm, tiptoeing on the suction cups. I saw a Caribbean pygmy octopus do this to catch a hermit crab part way out of its shell. The slow, smooth octopus approach was almost imperceptible, as I watched the flick of the antennae of the unconcerned crab. Then with a quick all-arms webover, the prey was snatched.

—Jennifer A. Mather

---

ingly, the eight arms can run into trouble, especially as they are poked into holes and crevices that might be occupied by an animal with sharp teeth. Octopuses share with lizards and brittle stars the ability to cast off an arm at a structural weak point and regrow that arm. Sometimes as much as half of an octopus population may have arm loss, but since the arms can regenerate before reproduction, this doesn't matter too much over the life span.

Mathilde Lange described this process of regeneration in 1921. At the stump of a cut-off octopus limb, bleeding is minimal at first, probably because the animal has direct nervous control of blood vessels and can close off the artery. After a while, blood is released for clotting at the wound, and the last pair of suckers is pulled toward the arm to minimize tension. There is then degeneration of both muscle connective and nervous tissue at the wound end of the arm before regrowth. A knob forms on the stump, and then over a few weeks, a new narrow arm begins to extend from it, developing new tissue. Chromatophores on the new skin come last. It's a good thing that the coordination of arms is flexible, since the octopus can lose several at different times, and arms grow back slowly.

The process of octopus limb coordination is not well understood. Each arm has a lot of independence, and while arm nerve cords come out from the subesophageal area of the brain, they are all linked by a circular nerve cord above the arms. No study of neural control and coordination of multiarm movement has been done since the work by W. J. ten Cate in 1928.

Still, octopus arm movement looks pretty uncoordinated. This flexibility may be because water offers a lot of support for an octopus, so it doesn't have to worry much about balance, but it also may be because octopuses don't always move forward. With lateral eyes and all those multipurpose arms, octopuses can and do move sideways. Several years ago, we had a student study an octopus walking. He found that the octopus usually walked on arms 3 and 4 of both sides but not always, and that there was no strict sequence of arm use. As in many other aspects of their lives, octopuses' walking is variable, apparently fitting the demands of the moment.

Octopus arms have a lot of jobs to do at once, from walking and feeling on the bottom to capturing and holding prey, grooming, and, for males, passing spermatophores. Unlike the highly specialized crab claws, legs, and mouthparts, physically octopus limbs are generalists, with only the male third right arm structurally different for sending spermatophores down its length to a female. The first pair of arms is generally used for exploration, and sometimes there's a size difference among the arm pairs, as in the Atlantic long-arm octopus, whose anterior arms are much smaller. But arm pairs usually have the same basic capacity.

While each arm can do many tasks, some arms actually perform a task more often. Ruth Byrne et al. (2006) looked further into such specialization in octopus arm use. When reaching out for food or a toy, the common octopus mostly used arms 1 and 2 of either side, but each octopus had a favorite arm of these four, not the same one for each animal. When an octopus needed more than one arm for a task, it recruited the ones next to the arm already in use. If you stimulate an arm nerve cord, the circular nerve cord linking arm to arm passes on the signal and the adjacent arms come into action. We call this the principle of neighborliness: using whichever arm is nearest.

There is a cost to having these multipurpose and multiunit arms. It requires extensive neural programming to set up all those contractions for the muscular hydrostat–based movements. Over half the neurons in the octopus body, therefore, are outside the brain, most of them in the arms controlling all these muscles. A nerve cord—a chain of ganglia—runs down the center of each arm. Below each of these many ganglia, a sucker ganglion controls the suction cup maneuvers. No wonder octopuses can weave and do pincer grasps and pass little crabs from sucker to sucker.

With all those neurons in the arms and only general central coordination of them, control of arm actions must be mostly local. Reflexes or local circuits control movement of these arm units, correlate a move with touch and chemical information, and even coordinate movement of nearby arm areas. This subject was last studied by C. H. Fraser Rowell in the 1960s. When physically threatened, several species, including the pygmy octopus, will cast off an arm at its base, and the arm then wriggles off on its own. A detached octopus arm can reach out for an object, withdraw from potential harm, and even act like a conveyer belt with a small food item. In the lab, isolated octopus arms (separated from the body) stimulated by electrical currents reached out in a smooth, bending wave from base to tip. Since it was the same motion as for intact octopuses, this organization of reaching seemed to be under local and not central control. That's a lot of decentralization, and the hierarchy of actions and reactions in the octopus needs more study.

Is there a disadvantage to this decentralization? Maybe without central monitoring of position, an animal can't sense its own body position. Wells set up a situation in which octopuses saw a crab through a glass window and then had to detour down a corridor to catch it. He concluded from his study that octopuses didn't know their location in space. We know now that this isn't true for the whole animal, as later studies on spatial memory and returning to home make it clear that the octopus knows where it is in the landscape. But in the smaller sense, what about knowing where all the parts of your body are and how big you are? It would be interesting to know whether a motor system that allows a set of suckers to untie surgical silk might have the down side of not letting the controlling brain know exactly what those suckers, and maybe other parts of the arms, are doing. While this way of controlling a complicated body isn't like the narrow centralizations in a big brain of mammals that we think of as connected to high intelligence, it certainly is a workable system for the octopus.

6
Appearances

Octopuses are far better camouflage artists than chameleons. They can change their entire appearance in less than a second. They can take on or get rid of fine gradations of color. They can make local changes like spots, bars, eye rings, and dark mantle edges. In addition to color alterations, they can assume skin texture changes and postures that mimic such things as smooth gravel, wavy curling seaweed, or other animal forms.

When octopuses evolved away from having a protective outer mollusk shell, their newly exposed body surface had a unique and complicated repertoire that allowed them to put patterns on themselves at will. Within the outer layers of octopus skin are many chromatophores—sacs that contain yellow, red, or brown pigment within an elastic container. When a set of muscles pulls a chromatophore sac out to make it bigger, its color is allowed to show. When the muscles relax, the elastic cover shrinks the sac and the color seems to vanish. A nerve connects to each set of chromatophore muscles, so that nervous signals from the brain can cause an overall change in color in less than 100 milliseconds at any point on the body, although local small areas called fields and ridges tend to color alike as units. When chromatophores are contracted, there is another color-producing surface beneath them. A layer of reflecting cells, white leucophores or green iridophores depending on the area of the body, produce color in a different way: Like a hummingbird's feathers, which only reflect color at a specific angle, these cells have no pigment themselves but reflect all or some of the colors in the environment back to the observer (see Messenger 2001).

The octopus's skin itself helps the animal make different appearances. There are small muscles within the skin that can pull it up into little peaks, papillae, to make the surface appear rough or smooth, depending on what texture the octopus wants to assume. These skin peaks are often largest above the eyes, and many photographs of octopuses show skin horns above the eye bulb that mask the appearance of head and eyes. Since octopuses have no bones and can put their arms into a variety of positions, these skin changes can also help the octopus to match its background. Young octopuses, in particular, can position their arms in fantastic twists and coils, including a set of postures labeled Flamboyant that make an octopus look like seaweed. These skin behaviors truly hide the octopus against its background. For volunteers watching common octopuses in Bermuda, learning how to find and follow the animals who were out hunting was among the more difficult things to learn. We told the observers that if they were looking directly and constantly at a foraging octopus and it apparently disappeared, they shouldn't worry. It was just doing a particularly neat piece of background matching, and if they just kept looking it would reappear when it moved.

While the colors and textures of the octopus skin probably evolved as camouflage, the ability to change them was later available for other uses. The evolution of the bony fishes in the marine environment challenged all the inhabitants of the seas. Only the numerous coleoids and a few nautiloids survived the competition in evolutionary terms. The coleoids did so by gaining jet propulsion, fast-tuned physiology, a skin display system, and intelligence. But researchers wondered: why would an animal that's colorblind evolve a color-displaying skin system? Fish rather than other octopuses were the designers of the cephalopod skin; octopuses that failed in ability to be invisible got eaten and removed from the gene pool. The colors that octopuses produce were aimed at preying fish, which, unlike the cephalopods themselves, can see colors.

Cephalopod camouflage is actually much more than simple background matching, although it works often enough for an animal to avoid being eaten. Hugh Cott wrote in 1940 about animal visual camouflage in all its variety, and discussed camouflage as simply not looking like an animal, noting that it's not necessary to look exactly like the background as long as you don't look like a recognizable form. This principle of nonform was also used by humans in disruptive camouflage during warfare, in weirdly painted clothes, and tanks or navy ships with wavy stripes of olive and blue and gray. The ships didn't look like the water around them to an observer in a reconnaissance plane, but they didn't look like ships either.

Background-matching camouflage is very common in marine animals. The peacock flounder is a dappled gray that blends with a sandy bottom perfectly, and it can enhance the match by flapping its body to dig into the sand and get loosely covered by it. Another shell-less mollusk, the sea hare (*Aplysia californica*), is a splotched pale brown with projecting wavy bits that mimic brown seaweed. Anglerfish are bumpy and gray-brown, blending with the rocks so well that unwary potential prey will swim right up to their open mouth and get eaten. Even the sea lemon nudibranch (*Archidoris montereysis*), whose clear yellow ought to be easy to see, lives on and eats yellow sponges, with which it blends perfectly. The abilities of octopuses, however, are special because they can use a lot of different background matches and change very fast.

Despite the variety of camouflage techniques available to them, octopuses often use background matching. A foraging common octopus in 3 ft. (1 m) of water would be highly visible if it didn't use camouflage. While crawling on a sandy patch and extending arm tips into the sand to explore for clams, it puts small pinpoints of paler and darker gray on its skin. Like the flounder, it flattens slightly and smoothes the skin surface so that it looks just like the sand bottom. Then, moving onto rocks with brown and greenish algae and snaking its arms between the fronds to find small crabs, it changes immediately to larger patches of mid-brown, olive brown, and greenish, more similar to the sea hare (see plate 21). The skin rises from a vague roughness of the surface to small papillae ranging from less than an inch (1 to 2 cm) in height and larger over the now raised head. And moving into the shade of larger rocks where fragile scallops hide, the common octopus becomes a rich plum-purple that matches the dark of the rocks, and its skin texture becomes smooth again.

Disruptive coloration is different from background matching; it can be used over the whole body or in specific surface areas of the cephalopod. The chunky little flamboyant cuttlefish (*Metasepia pfefferi*) can put wide bands of black, cream, and warm brown irregularly across its skin (see plate 22). The common cuttlefish (*Sepia officinalis*) can take on a similar kind of deception temporarily. On a rocky habitat, it can assume blotchy patterns of deep brown, tan, and white that match neither the colors nor the texture of the rocks but do make the animal disappear from our perception of an animal. This camouflaging method is a kind of distraction—making some part of itself obvious so that the predator's eye doesn't recognize the animal's form. The common octopus has a couple of white dots on its mantle that can accomplish this distraction.

Disruptive camouflage can be area-specific too. Octopuses especially need to disguise their head. For many animals of different groups, an eye is recognizably an eye and indicates an animal around it. Many animals have ways to disguise their eyes or to make fake eyes at the wrong end of their bodies. The four-eyed butterfly fish has a vertical black bar through its real eye and a wide round black eyespot at the posterior end of its body. A larger fish will dart at the apparent eye, the butterfly fish jerks forward, and the predator only gets a mouthful of water (see plate 23).

One way that octopuses disguise the slit pupil of the eye is to put an eye bar on the skin of either side measuring about an inch (2.5 cm), changing the eye's appearance so to an observer the eye is not round but just an extended line. The ability of the octopus to raise the skin papillae as horns above the eye's circumference also breaks up the eye outline so it seems to be irregularly wavy instead of circular. And they can extend the skin camouflage of the irregular blobs of color across all the eye skin except the pupil, so the eye seems to be just more skin with a line through it.

The importance of the appearance of the eye region for an octopus is indicated by the amount of neural programming devoted to it. Nine nerves control the expansion of the chromatophores that make the color patterns in all of the skin, but four of these come to just the small eye area. We obtained a clue as to how the octopus can control this appearance when we were studying octopuses' recognition of individual humans. After a few trials, when the person designated as the hassler approached, the octopuses put on eye bars, and when the one who fed them approached, they did not. We are now studying what changes constitute an eye bar and how its appearance can be conditioned.

Roger Hanlon et al. (1999) suggested that octopuses use their quick-change ability to cause even further confusion in predators. Any single altered pattern may not be especially deceptive, but an octopus discovered by a diver or a fish will puzzle the potential predator by using its sequential quick-change abilities. Fish have good visual memory and are likely to seek out prey by remembering what they look like—this is a "search image." The fish forms a visual search image of the octopus, but can't follow the prey after it changes, and will then lose sight of it. Imagine the challenge to the predator when a common octopus goes from a green and black background match to pale brown as it lifts off the rocky surface, then jets away and assumes stripes or gradually changes to a dark gray that spreads from the arms up over the body, then, settling quickly on the rock again, it becomes green and black by background matching.

The octopus's ability to take on skin patterns is also applied in situations other than camouflage. One dazzling example, which is displayed not

Quick-Change Artists

In avoiding predators, a cephalopod can pull amazing quick-change tricks like those Roland and I observed in the Hawaiian bobtail squid (*Euprymna scolopes*). When we tried to catch it, its avoidance techniques were awesome. Like an octopus, it turned from pale to dark, blew out a blob of dark ink, turned pale again, and moved away while we were still grasping at the dark ink patch. It then went to the surface and put on a Flamboyant pattern, looking like a piece of seaweed, and then went to the bottom to match the rocks below. A predator logically expects an animal to keep looking the same. The color-based evasion technique makes the prey not just hard to find but, by changing appearance unpredictably, impossible to follow.

—Jennifer A. Mather

to predators but to potential prey, is the Passing Cloud, named for the shadow of a cloud passing over the landscape. Early reports of the appearance of this feature were simply that it was a dark form on the dorsal surface, passing generally forward over the body, but the display was little known since it rarely occurred in the lab.

While studying the effect of food supply on octopus foraging in Hawaii, we kept several day octopuses in an outdoor saltwater pond on Coconut Island. We had a unique chance to watch and videotape behaviors that hadn't yet been described in detail. Back in the lab and replaying the video frame by frame, we found how complex the Passing Cloud display is. The Passing Cloud formed on the posterior mantle, flowed forward past the head and became more of a bar in shape, then condensed into a small blob below the head. The shape then enlarged and moved out onto the outstretched mantle, flowing off the anterior margin and disappearing. The process took less than a second. Since the display is usually used when the octopus is hunting and comes right after an attempt to surround a prey with the spread arms and web, a startle attempt is a likely explanation. Andrew Packard and G. D. Sanders suggested in 1971 that the octopus is conveying, "Move, you animal!" Any sudden movement will startle a crab that has frozen to evade being caught by the octopus. The crab's subsequent motion allows the octopus to notice and catch it.

The details of Passing Cloud displays are easier to understand as communication if you know the neurophysiology behind visual perception. The day octopus also puts an edge of white skin beside or behind the dark blob of the Passing Cloud, which greatly increases the visibility of the cloud shape by tapping into a feature of visual system analysis known as lateral inhibition. Organisms pay attention to change, not stability. No animal cares that the tree is still a tree, but a moving dark edge is likely trouble. By the lateral inhibition process, when receptors in the eye fire, they inhibit their neighbors. When a surface is uniform, that means the receptors cancel each other out and there's no signal to the brain. Edges, on the other hand, don't cause this kind of inhibition and are exaggerated. Darkness next to a bright area is construed as more dark. The day octopus may have evolved the white edges to make the dark cloud most visible to the crabs, which are the common prey of this species.

Another way in which the Passing Cloud production taps into visual physiology is more a matter of perception than physiology, because it produces apparent but not real movement. It's easy to say the cloud shape moves, but in fact no part of the skin or the octopus moves at all. What actually happens is a complicated series of chromatophore expansions and contractions across different skin areas. This phenomenon of seeing movement that isn't really there is called apparent motion, and it's used in light advertisement displays to humans in cities everywhere. If a sequence of dots appears close to each other in space and time, the brain simply constructs them as one moving dot. Making apparent but not real motion is useful to the octopus's own visual processing. Presumably it is looking at where the crab was before it froze and seemed to disappear. If the octopus itself moves to startle the crab, the image of the crab will blur on the retinas of the octopus's eyes, and the now-moving crab will be more difficult to pick out. If the octopus can stay still yet startle the crab by sending a Passing Cloud along its body in apparent motion, it maximizes its own visual ability and has a much better chance of catching the crab if the crab moves. This is a neat perceptual trick.

Octopuses also use this intriguing skin pattern production ability to display to each other. Surprisingly, we hardly know the precise signals and sequence of communication that allow octopuses to find mates and reproduce. We do know that these animals are basically solitary and semelparous—reproducing only once at the end of their lifespan. They aren't concerned with one another until maturity, and large octopuses may even catch and eat small ones. To reproduce, they must find each other, and ascertain species, sex, and maturity of the other individual. The male must also establish whether the female is ready to mate and bring her into readiness if she is not, and females especially must overcome distrust. Solitary animals have to do this fast; they may not get another chance. Finding, identifying, and mating is a complex sequence, and visual information from the skin must be useful only after animals have located each other, since it can be seen only at a short distance.

For years, Bill van Heukulem (1983) thought that the Hawaiian day octopus (sex unknown) signaled readiness to mate by taking on a pattern of wide chocolate brown and cream stripes that extended over the body and arms. But further observation suggested that the stripe pattern was a warning coloration, more evasive and challenging than sexual. At the pond in Hawaii, we saw mature male octopuses giving another distinct coloration when approaching a female. We named it White Papillae. The skin surface was raised in large papillae all over, the papillae were bright white, and the background skin surface was a deep brown, almost black. Since the octopus is color blind, the mottled black and white display is an easily visible high-contrast signal to the female that an interested male is approaching (see plate 24).

Through the accident of confinement, we learned about what might be referred to as the principle of spatial discouragement. In nonsocial animals widely spread out across space, the opportunities for mating are probably few, and octopuses, particularly males, must quickly take advantage of every reproductive opportunity. In our study in Hawaii, we kept two pairs of day octopuses in the pond for ten days, one pair after the other. Shortly after finding the female, each male spread White Paps gloriously over the skin surface and moved toward her, with his third right arm advanced, ready to find her mantle cavity and pass spermatophores. After some evasion and challenge, each female accepted the male, and mating ensued for about an hour. During the following nine days, as we observed the octopuses hunting and eating, each male tried again and again to mate, and each female rejected him. As in many animals, the female octopus will become uninterested after a successful mating; she can store sperm for months and has no need for more from him. Gradually, as the flashy show failed to win her cooperation, the male reduced the area of skin that displayed the White Paps. First it was only the side toward her; no need to waste a display on areas that she couldn't see anyway. Then it was only a few arms. Then finally, with rejection after rejection, such as moving off, jetting water at him, and even attacking, only his third right arm displayed the pattern. As his motivation level fell, so did the display, and after a week he quit trying and the pattern was gone.

Complex and varied as octopuses can be, all octopus species are not equally good at these visual skin displays. The shallow-water day octopus and the common octopus are the most studied octopuses, because humans are diurnal and prefer working in shallow tropical waters than in cold dark depths. Night active and deep-sea octopuses have a much smaller repertoire of visual displays. The displays of the small and strictly nocturnal Caribbean pygmy octopus range minimally from pale to blotchy to solid warm brown. The zebra octopus (*Octopus chierchiae*) has constant zebra stripes, the white-spotted night octopus (*O. macropus*) is reddish-brown but lines its arms with large white dots, and the deep-sea spoon-arm octopus is nearly a constant red brown. The deep-sea cephalopods have evolved their own ways to contact visual receivers such as other octopuses, predators, or prey. Many, like the vampire squid (*Vampyroteuthis infernalis*), have bioluminescent spots, and some even replace dark ink with jets of luminous bacteria.

One thing we know particularly little about is how the octopus controls all these color and texture changes, mostly because of our lack of research on the brains of these animals. For example, how does a color-insensitive animal know how to do color matches? We just don't know yet, and the problem is far from simple. An octopus receives visual information about its surroundings and situation and then computes their appearances. It must then choose, whether consciously or not, what to do about it. Then it turns on output circuitry that leads to the chromatophores that must be used, and to skin muscles and arm musculature, telling them to carry out the proper action to make texture. Before long, the octopus is also comparing and gauging results to decide what to do next.

Hanlon (2007) was particularly impressed when an octopus did the "moving rock" trick. Crawling across an open sandy area (and thus vulnerable), it went gray-green with a lot of raised papillae, looking much like a rock covered by a clump of seaweed. Instead of walking normally, it drifted irregularly across the sand, swaying slightly as seaweed would if pushed by currents.

We also need to do more study on where and how the brain of the octopus matches and programs the dazzling variety of patterns it assumes. A chromatophore lobe in the subesophageal part of the octopus brain controls the display system itself, but it's the last step in output, and many of the circuits that control specific changes must be far from there, possibly in the optic lobe with its intimate connection to the eye. Learned visual information is stored in the vertical lobe, and this too must be used if identity of critical stimuli or appropriate responses is learned, which means that all the brain must cooperate, but we don't know exactly how. Roddy Williamson and Abdesslam Chrachri pointed out in 2004 that this interesting neural network has complex descending control but probably no feedback loop. They believe that the octopus doesn't monitor the patterns it's displaying, such as white spots (see plate 25).

Brain control circuits for octopus skin displays, wherever they are, may be of great variety. Any species-typical displays, such as White Paps, must be fairly fixed, because sexual displays need to be clear. If you have only a few chances in your lifetime to seduce a mate, the signals need to be produced without much learning. They also need to be simple and easily recognizable, if every member of the opposite sex is to quickly recognize what each display means.

But the circuitry for camouflage includes commands for posture and skin texture as well as pattern, so camouflage output must be more variable and tied quite directly to perception. The circuitry to control the Passing Cloud display might be fairly simple. Still, the Cloud must "move" toward the location of the prey when an octopus directs a startle display at a particular crab, making the possible programming for controlling chromatophore muscle contraction more complex. Perhaps the biggest challenge in imagining the octopus brain circuitry needed to control appearance is related to changing displays. As Hanlon et al. pointed out (1999), the octopus must be able to choose and produce a display, move, change, measure the results, and make a decision as to which one to make and what to do next. This set of actions is controlled by far more than fixed neural circuits. Perhaps this is what octopus intelligence does best, producing the flexibility and variety in control of a system whose functioning we barely understand yet.

7
Not Getting Eaten

Cephalopods such as octopuses spend much of their lives working on defense. They have to be constantly vigilant about predators to whom they are a neat unprotected package of protein. Having given up the external shell of their mollusk ancestors, they have to rely on hiding, camouflage, and the intelligence to choose different tactics to avoid getting eaten.

Among the cephalopods, vulnerable octopuses can hide in a den, and juvenile cuttlefish find refuge by burying themselves in the sand, but others such as squid never rest during their waking hours. Sleep has not been reported in squid and only occasionally in buried juvenile cuttlefish, and it has only recently been seen in octopuses. We have spent many hours observing Caribbean reef squid and have never seen any sign of rest. In fact, they appear to be the Nervous Nellies of the cephalopod world: squid schools spook eight times per hour when resting in the open. It's possible that squid sleep with half their brain at a time, like migrating birds and marine mammals, but we have seen no indication of that.

Many ocean animals prey on subadult or adult octopuses. Small octopuses are eaten by small fish and crabs, and most fish would not pass up an octopus away from its den. In the tropics, often-present fish such as grouper and barracuda eat octopuses, and in the North Pacific, lingcod and wolf eels will eat them. John Randall (1967) has pointed out that Caribbean predatory fishes, such as squirrelfish and yellowtail snapper (see plate 26), are generalists: they eat most any fish that will fit into their mouths and many invertebrates including cephalopods. In the 1960s, he discovered this by taking samples of many Caribbean fish by trawling, by hook-and-line fishing, and with the help of narcotizing drugs. He then cut open their stomachs and examined the contents to see what the fish had been eating. He probably wouldn't be able to do such a study nowadays, as many fish stocks are depleted and some species he took are now rare or endangered, so we are grateful for his older, thorough study.

Sharks, especially the many species of dogfish sharks, are octopus predators. Based on the stomach contents of sharks, researchers have established that in shallow tropical waters, octopuses are a big part of the diet of white-tip reef sharks and nurse sharks. And in the deep sea, a number of other deep-water sharks eat octopuses.

Among the best known and most persistent of octopus predators is the moray eel. Because of its body design, it can snake into small areas such as octopus dens for prey, and it also is a major predator of octopus in shallow tropical waters. Moray eels can go through small openings under rocks looking for small fish and crustaceans, and they frequently find an octopus. Morays can sense an octopus chemically. Octopuses and moray eels are about the same size, so when they meet, a battle may ensue, each fighting for its life. An octopus will also readily eat a moray eel. The moray sometimes wins, but many times it only twists off an arm of the octopus. The octopus—now with only seven arms—jets away, still able to hunt its own prey while the arm regrows. In California, scorpionfish and two-spot octopuses have a similar relationship: large scorpionfish eat small octopuses and large octopuses eat small scorpionfish.

We made a discovery about this kind of predator-prey relationship at the Seattle Aquarium. The largest tank there is a multispecies tank of 400,000 gal. (1,560,000 l). Giant Pacific octopuses used to be kept in that tank, several at a time, so that at least one would likely be visible to the public. The tank had a mix of large northeastern Pacific fish, including wolf eels and dogfish sharks about 3 ft. (1 m) long. Since dogfish are known octopus predators, we were a bit leery of putting them together. The public usually doesn't like to see predation in action; we didn't want the dogfish biting arms off the octopus in front of visitors. Things seemed to go well for a while, then something surprising happened: some of the fish, including dogfish, started disappearing. Partially eaten dogfish carcasses were found in the morning on the bottom of the tank. We tracked down the culprit, a 60-lb. (27-kg) male giant Pacific octopus. We eventually were able to videotape this behavior: he would follow the rather slow-moving dogfish, catch it, and eat it. As in the case of the scorpionfish in California, dogfish eat small giant Pacific octopuses and, at least in the confinement of a tank at the Seattle Aquarium, large octopuses eat dogfish.

In Puget Sound, there appears to be competition for den space between octopuses and wolf eels. Wolf eels are wolf fishes, not true eels, that grow to nearly 8 ft. (3 m) in length and weigh about 50 lb. (23 kg). They have large canine and molar teeth that they use to crush crabs and sea urchins, ignoring the spines piercing their lips. They live in the same caves and crevices that octopuses use and are found around the rim of the North Pacific. Wolf eels are known to eat octopuses, and maybe octopuses eat wolf eels. The giant Pacific octopus gets big, so in this wolf eel–octopus competition, the octopuses usually win the battle over choice dens and also for their lives.

Though we were wary of the octopus–wolf eel interaction, at the Seattle Aquarium we decided to have a multispecies exhibit that contained a giant Pacific octopus and a pair of wolf eels in a 3000-gal. (12,000-l) tank. We designed the tank decor to exhibit both species together by providing a number of clefts, cracks, and overhangs in the rockwork for both the octopus and the wolf eels. When the exhibit opened, we put the wolf eels in the exhibit first and the octopus a day later. The two species immediately took up residence in a different area of the exhibit, showing that they can live in the same area and not kill each other as long as there is enough shelter. Several months later, the 40-lb. (18-kg) octopus was moved out of this exhibit and a smaller one of 15 lb. (7 kg) was moved in. This octopus moved directly into the large male wolf eel's cave and displaced it for two days without any overt sign of aggression; it seemed to have a bold personality. After two days, the wolf eel took back its cave. There may have been wolf eel–octopus wars going on at night for the possession of this prime den, but we don't know.

Other main predators of octopuses are marine mammals, particularly seals. In the North Pacific, sea otters are major predators of the giant Pacific octopus. In Alaska's Prince William Sound, David Scheel et al. (2002, 2007) studied the return of giant Pacific octopuses and sea otters after the Valdez oil spill disaster of 1989. He found that octopuses were using the shallows and intertidal zone and also the area deeper than 100 ft. (30 m), areas above and below the sea otters' usual foraging range. He speculated that the sea otters had simply eaten the octopuses between those two depths. Of course, river otters are casual visitors to the marine world, and in coastal areas they spend considerable time on shore and in the shallows. They, too, have been reported to catch octopuses, so the octopus isn't safe in the intertidal region.

Many sea birds and shore birds will eat octopuses but not as a major part of their diet. Crabs and gulls are thorough foragers in the intertidal zone, so an octopus in a tide pool is at risk. Sea birds such as penguins eat large quantities of squid, and sometimes cephalopods eat birds, as well. In Puget Sound, giant Pacific octopuses have been seen eating seabirds—gulls and alcids. At low tide, one octopus with a den at the bottom of an intertidal boat launch ramp on Whidbey Island caught and ate both glaucous-winged gulls and pigeon guillemots.

In some cases, supposed octopus predators are probably dining on senescent octopuses, particularly males that have mated already or females whose eggs have hatched. Scavengers along the beach, such as wolves and coyotes, are often blamed for the deaths of octopuses they are feeding on. At the end of their life cycle, octopuses don't act normally—they don't eat and they don't live in dens—so they are extremely susceptible to predation. Even among fish, when stomachs of supposed predators are examined and octopuses are found, it is quite possible that these fish were simply preying on dying, senescent octopuses rather than actively hunting down and eating active, alert octopuses.

It may be surprising to learn that 20-ft. (7-m) killer whales, or orcas, eat giant Pacific octopuses. These orcas live in the North Pacific; they are large, the largest of the dolphins, and can select large prey. Maybe the octopuses they caught were senescent and were crawling or swimming outside of their dens, where they were highly susceptible to being eaten, because it's difficult to believe that an orca could capture an octopus inside a den or a healthy octopus out hunting at night on the bottom.

A recent observation by scuba divers proves the susceptibility of old octopuses to predation when they are exposed. Scuba divers reported seeing three large male giant Pacific octopuses out in the open at a popular dive site in Puget Sound. They estimated that each animal weighed over 60 lb. (27 kg), a normal size for adult male giant Pacific octopuses. Each was missing at least four arms, and one was missing at least half of each of its arms and was crawling around on the stubs. During this same dive, the divers saw a 12-ft. (4-m) six-gill shark make a pass over the octopuses. It is likely that these male octopuses senesced at about the same time, reaching the natural end of their lives, and ended up providing food for predators such as the shark, which was scavenging on the near-dead octopuses.

Although many predators such as sharks prefer to eat live prey, most are not averse to eating carrion like dead octopuses. Carrion—animals that have died from some other means—is a good source of food that doesn't fight back. Many animals we think of as fierce predators eat a lot of carrion. Lions much prefer to eat a dead herbivore on the African veldt than spend the considerable energy needed to stalk and catch it, running the risk of getting gored by horns or kicked by sharp hooves. The bald eagle, often envisioned as hunting down small mammals, actually gains much of its nourishment from salmon that die after they spawn in rivers.

The main refuge against predators for a shallow-water octopus is its den (see plate 27). Since octopuses live in relatively small dens, often not much larger than themselves, with a small or narrow opening, predators usually can't get to them through the small opening, capture them, and pull them out. We have observed that octopuses need a place of protection while they sleep, and they do this in dens. The octopuses that construct dens in sand or mud are also protected, because predators that would try digging them out would get a mouthful of mud.

If a shallow-water octopus is in its den and a moray eel, wolf eel, or invading octopus comes along, it has several methods to repel them. At rest, it sits in the opening of the den with one dorsal (first) arm across the opening with the suckers facing outward. It usually has one eye peering out over the arm. Incidentally, individual octopuses are either left eyed or right eyed, usually looking consistently with one eye or another, just as we humans are left handed or right handed. In this posture, the octopus can see the action going on outside the den, it can readily lash out one or more arms at the invader, and its mouth is positioned outward, ready to bite with the beak and inject deadly venom, if necessary, when attacked. Octopuses make this defense more effective by blocking up the entrance to their dens with rocks and shells, further preventing predators from seeing them or getting at them. Jennifer calls this construction of a defensive wall "tool use" by the octopus.

When a potential predator gets too close to an occupied octopus's den, the octopus may deter the predator by jetting a blast of water at it from its funnel. Octopuses mostly use this method to get rid of annoyances, such as nonthreatening fish like perch or rockfish in the northern Pacific, or parrotfish and wrasse in tropical waters. Humans trying to sample shells from the midden can also annoy octopuses and receive a water blast. When researchers were extracting giant Pacific octopuses from their dens for scientific study by squirting a noxious chemical into the dens, often the chemical was blown right back out at the researchers by the octopuses' water jets.

Octopuses jetting water at annoyances can lead to some amusing consequences. In the late 1950s, a researcher named Peter Dews wanted to conduct a standard learning experiment with an octopus, so he set up a situation in which three common octopuses had to pull a lever to get food. Two octopuses learned the process easily, but the third, named Charles, was a challenge. In Dews's words, "Charles had a high tendency to direct jets of water out of the tank; specifically they were in the direction of the experimenter." This octopus was either annoyed by the researcher or had a low tolerance for annoyances in general. Individual octopuses have different personalities, and octopuses with some temperaments just aren't suitable for this kind of experimentation. Charles eventually pulled the lever out of the tank wall.

At the Seattle Aquarium, there was a female octopus with an aggressive personality. This octopus would squirt water out of her tank at a particular employee, a night staffer, sending a large amount of water directly at the person each time she checked her during the night. The night staffer got soaked from this unexpected midnight shower, and became irate when she thought the octopus was singling her out for special humiliation. We eventually figured out that the staffer was shining her bright flashlight on the octopus tank to check the water inflow and outflow. The octopus rightly associated this midnight disturbance with the staffer and showed her annoyance by jetting at her.

We witnessed an amusing example of octopuses jetting at annoyances in Hawaii. We were observing a Hawaiian day octopus in an enclosed saltwater pond on Coconut Island. It sat on the edge of a concrete slab that had fallen to the bottom of the pond, just outside the den it had excavated under the slab. A butterfly winged erratically over the pond above the octopus. When it came close to the water surface, the octopus blew a sudden strong burst of water straight up through the 1-ft. (30-cm) depth of water. The water erupted like a geyser beneath the butterfly, which darted away quickly.

Since octopuses face many types of predators, they use a variety of means to avoid getting eaten while out of their dens looking for food or mates. Their first line of defense is not getting seen and therefore they use camouflage. The octopus species in shallow tropical waters may be the best color-change artists of the cephalopod world. They live on, work in, and camouflage to the multihued coral reefs. In northern waters, octopus species live in darker water frequently clouded with runoff sediments or plankton, and at those latitudes the sun is at a lower angle, so octopuses and fishes there specialize in grays, reds, and browns. Deeper in the ocean, along the continental shelf or into the abyssal plain, octopuses are almost uniformly muddy brown or gray because there is no light there. Deep-sea predators may still possess eyes only to detect bioluminescence from other animals.

If camouflage doesn't work and a potential predator gets closer, an octopus may go to a different strategy, still using a change in appearance. It may use a bluff behavior: it either makes itself look bigger than it actually is, or it may assume the appearance of something else entirely, something that isn't a tasty octopus. This mimicry, or deimatic display, in cephalopods is an example of a display that startles a predator and makes it hesitate before attacking. The deimatic display is remarkably similar in many types of cephalopods. Cuttlefish show it by turning pale and putting on two dark spots near the posterior end of their mantle. Caribbean reef squid show it with a varied number of dots.

Shallow-water octopuses also produce eye patterns, showing two dark areas on the pale background of their skin around the eye while spreading the arm web wide and paling it, making them look much bigger than they are and mimicking an animal with big eyes set wide apart. Some species have permanent eyelike colored spots, or ocelli, in the skin in front of and below their real eyes, which they can intensify to make them look brighter. Although the ocelli may be used for deimatic display, observers can also use them for species identification.

The deimatic display is similar in several shallow-water cephalopods. Since these ocelli are located on different areas of the skin in different cephalopods, the display probably evolved in parallel, with different animals arriving at the same solution to the same problem, or convergent evolution. Animals in other phyla, such as the four-eyed butterfly fish and the luna moth, use a similar eyespot display.

There is some debate as to how octopuses use deimatic behavior and display. Deimatic display among squid is used in cases of mild threat, such as a big parrotfish getting too close to a squid. No effect has yet been established for the use of deimatic display in octopuses. Studying antipredator behavior among octopuses is a challenge because scientists rarely get to see predation events. We've witnessed lots of evasion behavior, but there hasn't been much chance to see it fail.

In addition to camouflage, octopuses use other defense behaviors while out of the den. Hanlon et al. (1999) observed that when predators approached, day octopuses changed their body patterns three times a minute when they sensed a threat, using patterns that were at times cryptic, conspicuous, and even mimicking of fish. They speculated that such changeability in body patterning prevents predators from developing a search image for octopuses. One particularly fascinating behavior they saw was an octopus crawling across a sand flat from one coral reef to another. The octopus perfectly resembled either a coral rock or a ball of detached algae as it moved across the sand, tiptoeing on its suckers. It was conspicuous against the sand but camouflaged perfectly to resemble the growth on the coral rocks of the reef yards ahead of it.

If camouflage or startle behavior doesn't work in avoiding predators, an octopus has more evasion techniques at its disposal before it would have to fight for its life. If given the opportunity, it will try to flee rather than fight with a predator or larger octopus. Shallow-water octopuses have a fairly typical escape response when predators get too close. First they turn pale all over, eject a cloud of ink, and jet away. Then they settle onto a new spot and quickly camouflage to match it. An octopus may use only one of these actions, or, when under the most threat, it might do all of them in sequence.

The octopus's sequence of paling-inking-jetting-camouflage has a confounding effect on a potential predator. First, the octopus is gone or appears to have disappeared very quickly. In general, octopuses don't swim as well as squid, but they can swim quite fast over short distances using jet propulsion. They usually swim a relatively short distance, hopefully to a pile of rocks or coral, where they can crawl in and hide or quickly camouflage, or just out of a predator's sight.

Avoiding a predator by blowing a blob of ink in its direction can have several effects. Octopus ink is largely formed of melanin, one of the blackest, most opaque substances produced by animals. Cephalopod ink was used in ancient times as the first ink for writing, and sepia ink is named after the common cuttlefish, *Sepia* sp., that provided it. Ink was produced by the earliest of the octopuses' coleioid ancestors. The fossilized ink sac from a 65-million-year-old cuttlefish, when ground up and mixed with alcohol, still made very good writing ink.

Octopus ink can be different colors in different species. One way to distinguish between the two sibling species of two-spot octopuses in southern California is their ink: one produces black ink and one produces brown. Nocturnal species of octopuses and deep-water octopus species may produce a deep red ink. As seawater filters out red light, the red color of this ink is absorbed quickly in water, so red ink looks the same as black in the deep.

The ink sac of an octopus sits underneath its digestive gland. It is composed of a small gland that actually produces the ink and a larger sac where it is stored. The sac has an extended duct that leads to the anus of the octopus and vents into the funnel. The ink duct itself has several pouches where ink is stored, ready for quick discharge out the funnel with a vigorous water jet.

Along the way to the octopus's funnel, the ink passes through glands, where it is mixed with varying amounts of mucus. The mucus gives differing consistencies to the vented ink. The funnel ejects the ink in a blob of a certain form. Some cephalopods eject ink in a blob about the same size and shape as their bodies, and so they leave a phantom squid or octopus (a pseudomorph) hanging in the water as they turn pale and jet away. Hawaiian bobtail squid leave a phantom squid blob several times in succession, each about 1 yd. (1 m) apart. Sometimes a squid then turns dark and hangs in the water at the end of the string of ink blobs, further confusing the predator by imitating the blobs. In lab tests, we saw that both the bobtail squid and the stubby squid (*Rossia pacifica*) could vary the consistency of their ink blobs from a diffuse cloud to a thick blob. In the Caribbean, we have seen squid ink blobs drifting in the current that came from a school of squid hundreds of yards upcurrent, proving the ability of an ink blob to hold its shape long after it is formed.

Octopus ink can also be squirted out in a long string as the octopus swims away. Most octopuses squirt the ink out in a big loose cloud, with little or no mucus mixed in. In this form, it has two functions. The first is to act as a smokescreen to hide the octopus from the predator's view while it camouflages against the bottom or turns transparent, jets away, and then camouflages again. Such ink clouds can be very large, many times the volume of the octopus that created them. Even a common octopus or a day octopus can make an ink cloud large enough to screen from divers, and a 75-lb. (34-kg) giant Pacific octopus can expel enough ink to obscure vision throughout a 3000 gallon (12,000 l) tank.

In addition to blocking a predator's sight, cephalopod ink contains tyrosinase, a highly irritating substance that temporarily paralyzes the sense of smell of a predator and also irritates its eyes. The mucus in the ink can also clog a fish predator's mouth and gills. Ink is such an effective defense that almost all cephalopods possess ink sacs. There is fossil evidence that ancient ammonites, belemnites, and even some fossil nautiloids had them. All species in the genus *Octopus* have ink sacs. The modern cephalopods that don't have ink sacs probably lost them through evolution. Other groups of octopuses, mainly deep-water types such as the vent octopus, don't have ink; ink would be no use in the depths where there is no light to see. The velvet octopus (*Grimpella thaumastocher*), a shallow-water octopus, doesn't have ink. It may have evolved from deep-water species and then moved into shallow water. We tend to believe that deep-water octopus species evolved from shallow-water ones, but maybe not always.

Some octopuses and other cephalopods have adapted their use of ink. One species of deep-water squid, *Heteroteuthis dispar*, doesn't seem to have melanin in its ink but instead expels a bioluminescent cloud. Since it's not effective to use black ink in the dark depths, and these little creatures can't make light by themselves, they culture bioluminescent bacteria in their ink sacs for use when they want to make some light. It must be thoroughly astounding to see these squid responding to a threat by spewing out this light cloud in the depths.

Some open-ocean cephalopods are barely visible. The glass octopus normally has transparent body parts and is barely visible, giving us an idea of the invisibility of tiny octopus paralarvae. In the face of a threat such as an ROV, or submersible, with its bright lights, glass squid can turn themselves inside out, then blow their mantle cavity full of dark ink, and look like a black basketball floating in the dark water. Maybe they are presenting another form of camouflage and they just don't look like edible squid, or maybe a predator who tries to eat them ends up with a mouthful of noxious ink.

Another amazing use of ink can be found in the broadclub cuttlefish (*Sepia latimanus*). Cuttlefish are known for their ability to produce the Passing Cloud. Broadclub cuttlefish can move the dark pattern across their skin from the posterior to anterior of the animal. We have photographed a cuttlefish displaying a broad Passing Cloud from the posterior forward, culminating with a squirt of ink toward another cuttlefish. This passage of a dark blot on the animal to the water is another way to confuse a potential predator about where the animal actually is.

The mimic octopus displays other bizarre and enigmatic defensive behaviors. This species, which has only recently been described by taxonomists, lives in sand and muddy areas of shallow-water Indonesia. Its claim to fame is its supposed ability to mimic the shape and behavior of other animals—animals that are less likely to get eaten than a tasty octopus. This octopus can flatten its body and move across the sand, using its jet for propulsion and trailing its arms, with the same undulating motion as a flounder or sole. It can swim above the mud with its striped arms outspread, looking like a venomous lionfish or jellyfish. It can narrow the width of its combined slender body and arms to look like a striped sea snake. And it may be able to carry out other mimicries we have yet to see. Particularly impressive about the mimic octopus is that not only can it take on the appearance of another animal but it can also assume the behavior of that animal.

The mimic octopus has become well known because of its defensive behaviors. Photographs of this octopus have been published and its behavior captured on film for television nature shows. Few people have seen or photographed the mimic octopus, because it lives in such a remote location and because few divers dive on the mud flats where it lives. Mimic octopuses have proven impossible to keep in captivity, even though they occasionally show up in pet stores. Roy Caldwell (2000) presents a good case for not buying these fascinating creatures: he believes that overfishing for the aquarium trade may drive the species toward extinction.

We don't yet understand why the mimic octopus has such ostentatious behaviors. Scientists speculate that this octopus mimics other animals that a predator wouldn't want to eat—an aposematism, when an animal imitates or looks like another that tastes bad or is venomous. If the mimic octopus can resemble the poisonous Moses sole, a stinging jellyfish that no fish would want to eat, and even a frilly venomous lionfish, then surely the normal octopus predators, such as grouper or barracuda, would avoid it.

This apparent mimicry behavior is also common in other shallow-water octopuses and a few deep-water species of octopus. Several widely scattered tropical species can make the same body patterns as the mimic octopus. An undescribed species in Hawaii does it. We have seen the Atlantic long-arm octopus do it in the Caribbean waters of Bonaire: it flattened itself and cruised over the sand just like a sole. We shall no doubt discover more unusual behaviors among tropical octopuses as they are observed more in detail. The tropical, shallow-water, Indo-Pacific octopuses are particularly likely subjects for future discoveries.

If camouflage, hiding, frightening, inking, bluffing, or blowing water jets at a predator do not confuse or scare it off, the octopus may have to fight for its life. If the intelligent octopus's home range has many predators, it may move away to a different area, which isn't a problem for octopuses since they don't hold territories and don't spend much time in any one den.

A Bad Bite

Humans sometimes get bitten by octopuses, which is, after all, simply defense on the part of the octopus. One day, a Seattle Aquarium employee was leading a beach tour at Saltwater State Park just south of Seattle on Puget Sound. Nearby scuba divers had caught a red octopus, which they brought to shore and placed in a bucket to show to the group. As the aquarium employee handled the octopus properly with a gloved hand, it abruptly bit him on his wrist just above the glove. He didn't notice the bite at first because there was no pain.

About a minute later, he noticed blood and saw the bite wound. The small puncture was less than 1/8 in. (3 mm) wide, which is the size for a bite from a small octopus with a body 1 in. (2.5 cm) long. When he noticed the bite, he sucked on the wound to extract the venom, but this didn't do much good. After about ten minutes, the wound site began to hurt. When swelling and fiery pain began to extend up his forearm, he called the aquarium on his cellular phone.

Other aquarium employees suggested immersing the wound in hot water, as hot as he could stand. A nearby espresso stand supplied the hot water. About 20 minutes after the bite occurred, he poured this water directly over the wound and the adjacent area. The pain and swelling from the bite dissipated within a minute, but he still went to Harborview Hospital in Seattle, the area's main trauma care center. The hospital notified the aquarium to get advice on the treatment of an octopus bite, and a nurse then applied an ointment for the blisters caused by the hot water. The next day, the bite could scarcely be seen or felt, but the man had headaches and weakness for a week. This man was lucky. An untreated bite from a red octopus on an aquarium employee twenty-five years earlier left rotting flesh at the site for about a month and a deep scar.

—Roland C. Anderson

Jim Cosgrove and Neil McDaniel (2009) found that giant Pacific octopuses only live in any one den about a month before moving on. Octopuses do move on if predators are directly annoying them, or if a scientist forces them out with noxious chemicals. If an octopus survives an attack such as by a moray eel that comes right into its den, it doesn't go back to that den. The lack of attachment to a particular den is good for the octopus but hard on the researcher who wants to study the animal.

If an octopus is bitten by a predator or is forced to fight, it still has two methods of remaining alive—one passive and one quite aggressive. If an octopus is held by one of its arms, it may detach the attacked arm. If a fish or marine mammal bites off an arm, the octopus can swim away minus the arm.

Some species of octopus have a set area in their arms close to the body where there is a narrowing or stricture. These species—the banded string-arm octopus (*Ameloctopus litoralis*), for example—are relatively small and have long, snaky arms. They have developed the ability to autotomize, or cast off a limb when attacked by a predator. Supposedly, the predator will take the limb instead of the octopus, and will be kept occupied long enough for the octopus to escape.

If the octopus is facing a major threat from a predator, it can bite whatever is molesting it with its beak. A piercing bite from the hard, chitinous, parrotlike beak of an octopus can be a serious deterrent to predators. The beak is well-muscled and has flanges for muscle attachments and leverage. When an octopus bites because of a threat, it can also inject poisonous venom into the wound.

Octopuses don't always avoid being eaten, of course, especially by humans. For millennia, octopus has been a favored food among peoples of the world. And today, thousands of tons of octopuses are harvested each year for human consumption, primarily in Europe, Asia, and the tropical Indo-Pacific. So the octopus doesn't always win in the game of "eat or be eaten."

Plate 1. This view of *Octopus abaculus* reveals the eight arms, the head, and the saclike mantle.
Roy Caldwell.

Plate 2. This octopus, known as wunderpus (*Wunderpus photogenicus*), only recently named by scientists, shows that not all species use background-matching camouflage. This species is thought to mimic toxic or venomous creatures. Roy Caldwell.

Plate 3. The bright appearance of the circular areas on the skin of this blue-ringed octopus (*Hapalochlaena maculosa*) shows its warning coloration, which advertises that the animal has a deadly venomous bite. Roy Caldwell.

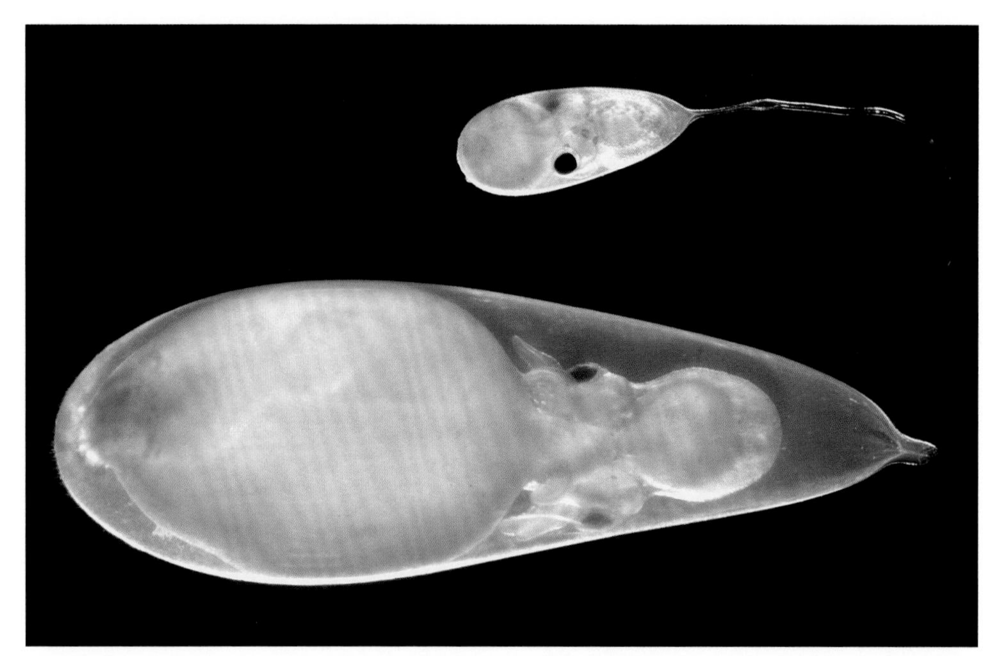

Plate 4. These small and large octopus eggs from the sibling species of two-spot octopus, Verrill's two-spot octopus (*Octopus bimaculatus*) above and Californian two-spot octopus (*O. bimaculoides*) below, give an idea of the species' relative size. John Forsythe.

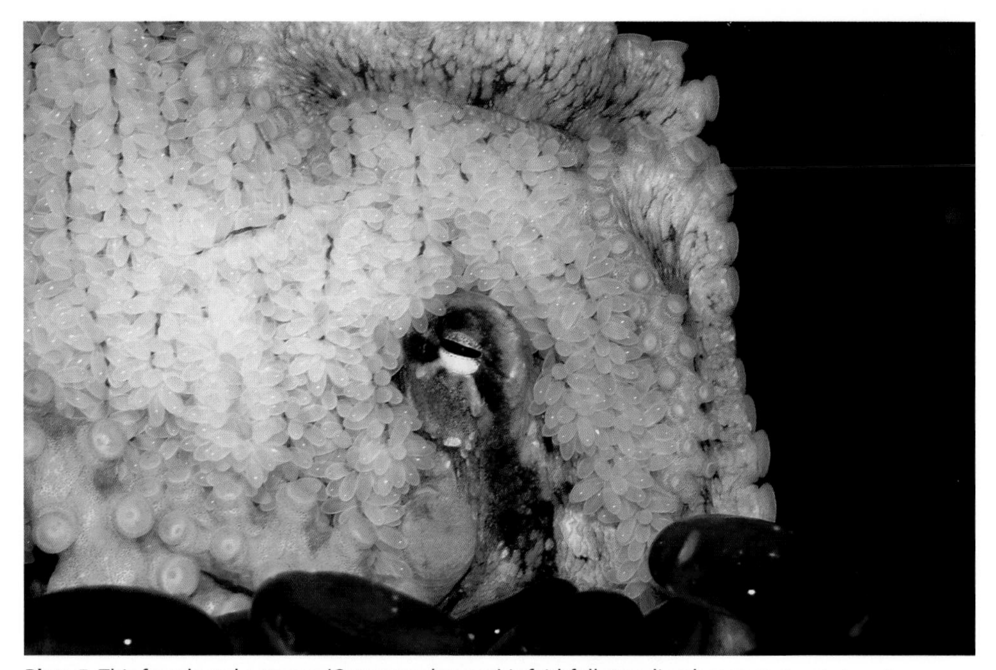

Plate 5. This female red octopus (*Octopus rubescens*) is faithfully tending her eggs. Seattle Aquarium.

Plate 6. Divers are in the water preparing to dive in the Seattle harbor. It is surprising to see an active dive site so close to the busy downtown. Seattle Aquarium.

Plate 7. Planktonic organisms come in a range of shapes and sizes. Seattle Aquarium.

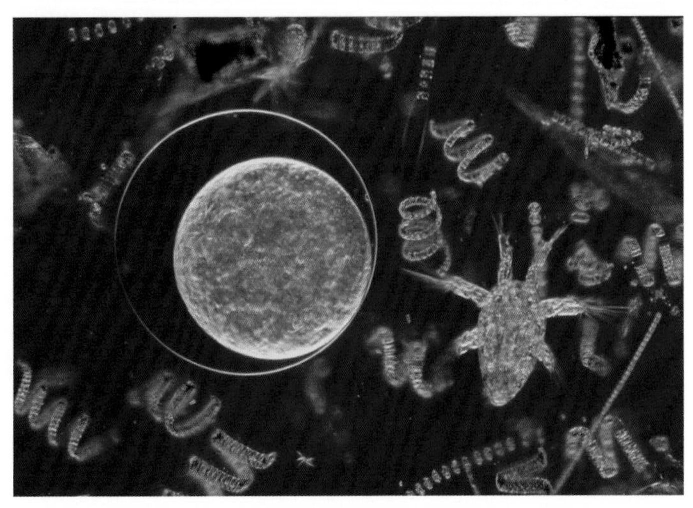

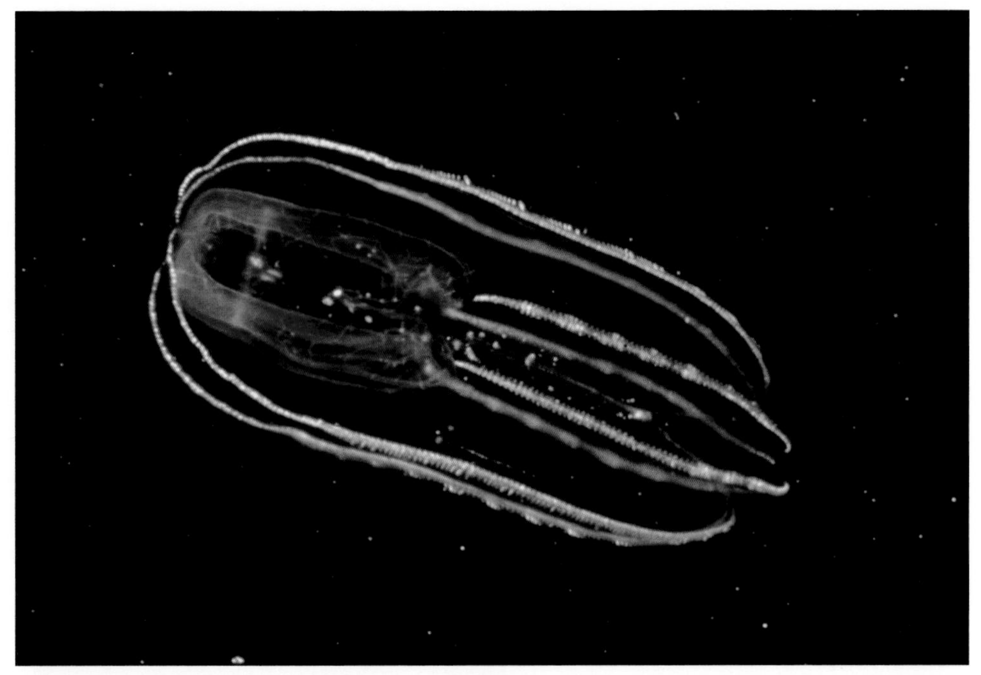

Plate 8. This comb jellyfish, or ctenophore, is one of the predators of the plankton, small to a human but huge to an octopod paralarva. National Oceanic and Atmospheric Administration, Department of Commerce.

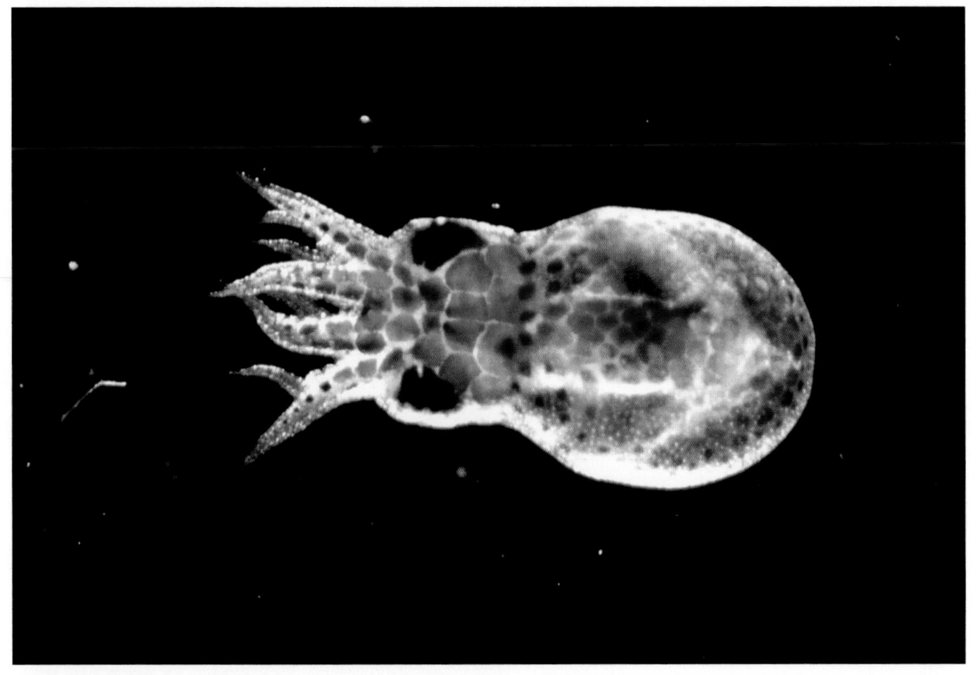

Plate 9. The body proportions of this paralarval Caribbean pygmy octopus (*Octopus joubini/mercatoris*) are clearly different from those of adult octopuses. John Forsythe.

Plate 10. This Caribbean reef octopus (*Octopus briareus*) is enveloping an area of the sea bottom with its spread arms and web. It will then explore the enclosed area with its outer arms for prey. Dry Tortugas National Park. James B. Wood.

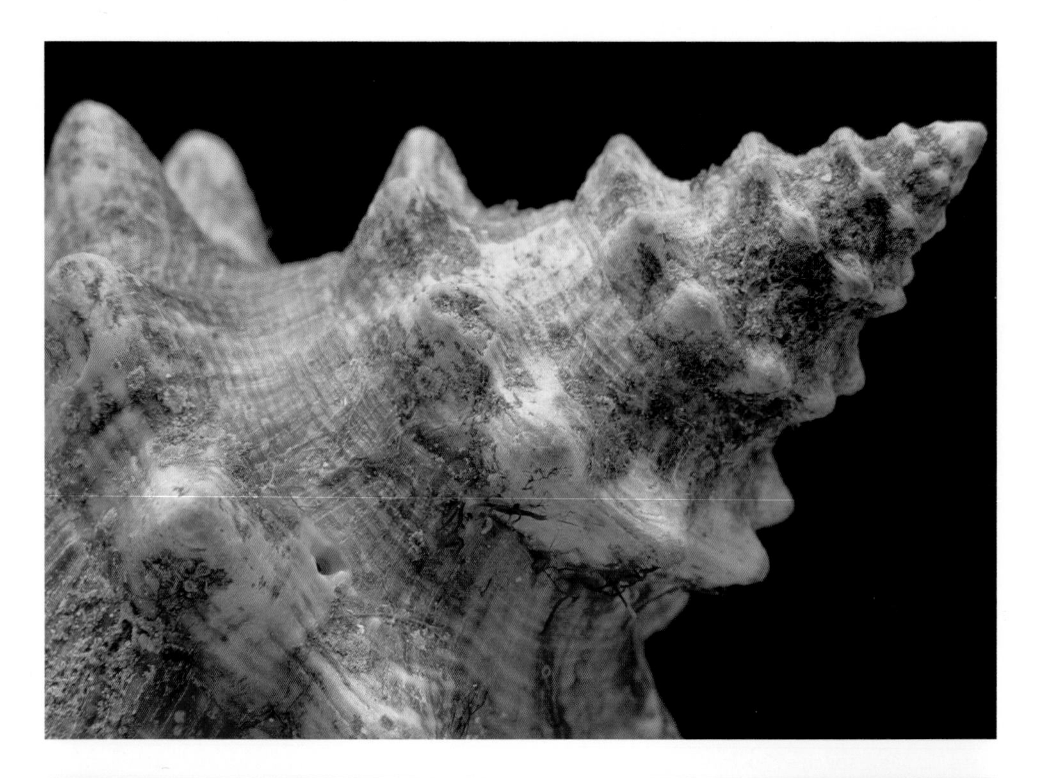

Plate 11. These two images show an octopus drill hole in the shell of a giant conch (*Strombus gigas*) from the Caribbean island of Bonaire. James B. Wood.

Plate 12. Some species of octopus, such as this wunderpus (*Wunderpus photogenicus*), burrow into and live in sand and mud. Roy Caldwell.

Plate 13. This common octopus (*Octopus vulgaris*) at the small island of Bonaire in the Caribbean Sea is well protected in its rocky crevice. It collected rocks to put in front of the den opening. James B. Wood.

Plate 14. These shell-less mollusks, lion nudibranchs (*Melibe leonina*), swim by bending their laterally flattened body and feed by scooping with their oral hood. Seattle Aquarium.

Plate 15. Scallops and this file clam (*Ctenoides* sp.) in Bonaire, an island in the Netherlands Antilles, use jet-propelled swimming by clapping shell valves together. James B. Wood.

Plate 16. A close look at this Caribbean reef squid (*Sepioteuthis sepioidea*) in Bonaire reveals the dual locomotion system—the funnel for jet propulsion (pointed downward, below the eye) and the lateral fins (along the sides) that undulate up and down. James B. Wood.

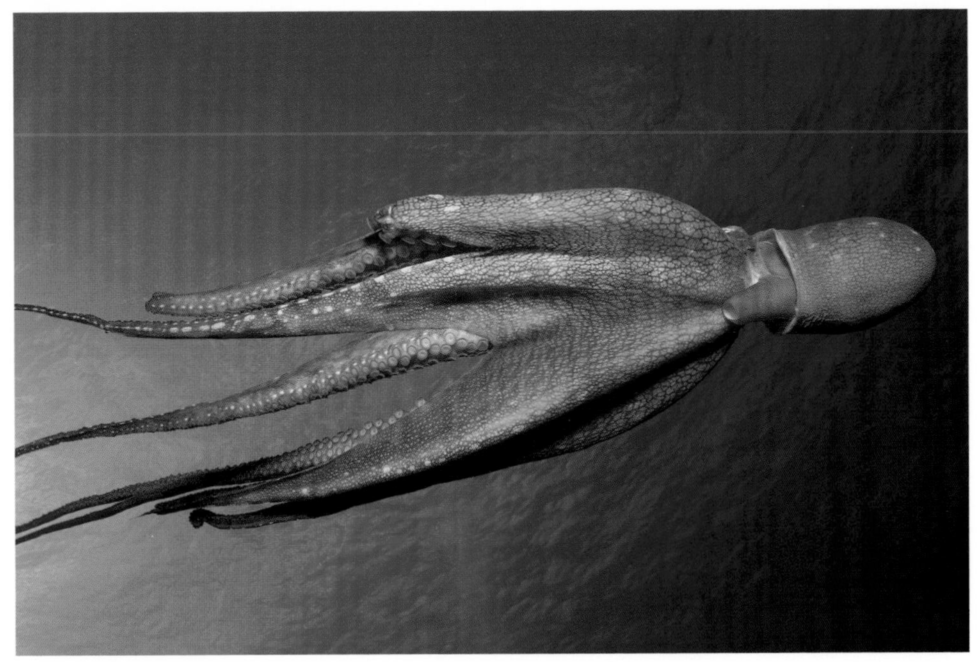

Plate 17. This giant Pacific octopus (*Enteroctopus dofleini*) has flattened itself in a gliding posture, which minimizes resistance and provides lift. © Stuart Westmorland 2009, http://www.stuartwestmorland.com.

Plate 18. By squeezing its arm until the diameter is very small, the Caribbean reef octopus (*Octopus briareus*) can extend the arm far out into the water. Dry Tortugas National Park, Florida. James B. Wood.

Plate 19. This common octopus (*Octopus vulgaris*) has curled its arms back toward its head and is rubbing the surfaces against each other to shed loose sucker skin. Captivity, Konrad Lorenz Institute, Vienna. James B. Wood.

Plate 20. Swimming crabs can be challenging for an octopus to catch, but the octopus can simply lift off the sea bottom and envelop the crab in a webover. Blue crab, Bermuda.

Plate 21. This common octopus (*Octopus vulgaris*) in Bonaire, Netherlands Antilles, has camouflaged itself well against the rocks. James B. Wood.

Plate 22. This flamboyant cuttlefish (*Metasepia pfefferi*) in Indonesia is displaying warning coloration. James B. Wood.

Plate 23. Fish have more than one strategy for not being seen and caught. The four-eyed butterfly fish, above, have their eyes concealed and a fake eyespot near the tail, and the scorpion fish, below, is well camouflaged. Bonaire. James B. Wood.

Plate 24. This male Hawaiian day octopus (*Octopus cyanea*) is displaying White Papillae, possibly a sexual signal or camouflage. John Forsythe.

Plate 25. The white spots on this octopus (species unknown) distract the observer not only from the real eye of the octopus (center top) but from recognizing the octopus as an animal. Roy Caldwell.

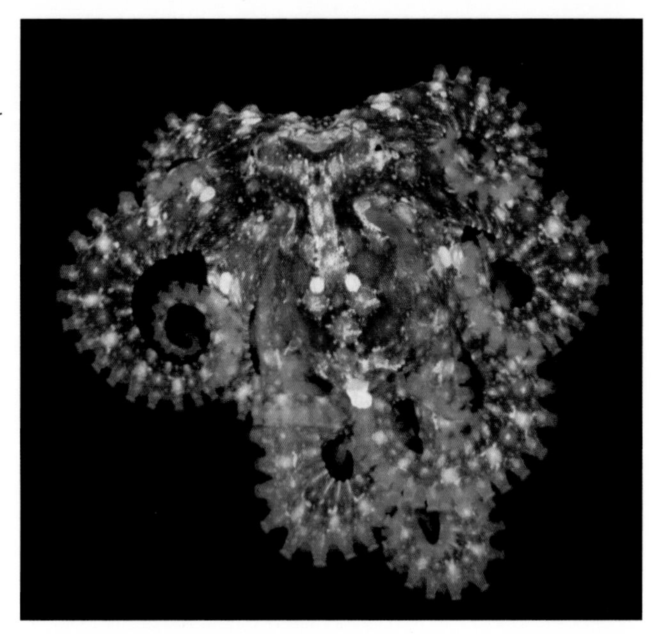

Plate 26. Most of the fish predators of octopuses, such as the Caribbean squirrelfish, above, and this yellowtail snapper in Bermuda, below, are opportunistic and will look for any unprotected animal to eat. Above: Abel Valdivia; below: James B. Wood.

Plate 27. Hiding is the first line of octopus defense. After retreating into the den, this common octopus (*Octopus vulgaris*) will pull rocks up to block the entrance, hiding it even more. Bonaire. James B. Wood.

Plate 28. This common octopus (*Octopus vulgaris*) in Bonaire could be described as bold—it is standing its ground and keeping its mantle inflated so it looks larger. James B. Wood.

Plate 29. This active male giant Pacific octopus (*Enteroctopus dofleini*) is not shy with aquarium guests. Seattle Aquarium.

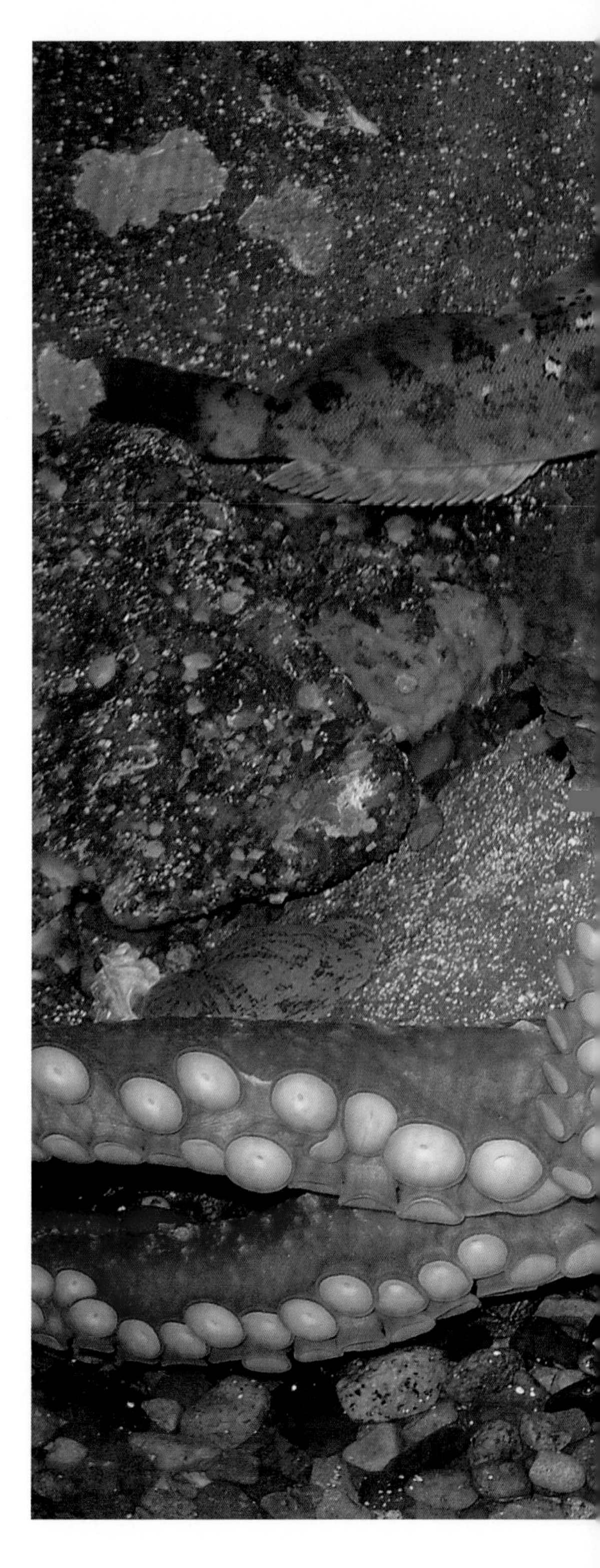

Plate 30. Foraging octopuses, such as this Caribbean reef octopus (*Octopus briareus*), keep track of where home is (spatial memory) and of where they are (working memory). Dry Tortugas National Park, Florida. James B. Wood.

Plate 31. This common octopus (*Octopus vulgaris*) kept in a lab in Vienna is investigating a glass jar with a lid. Konrad Lorenz Institute. James B. Wood.

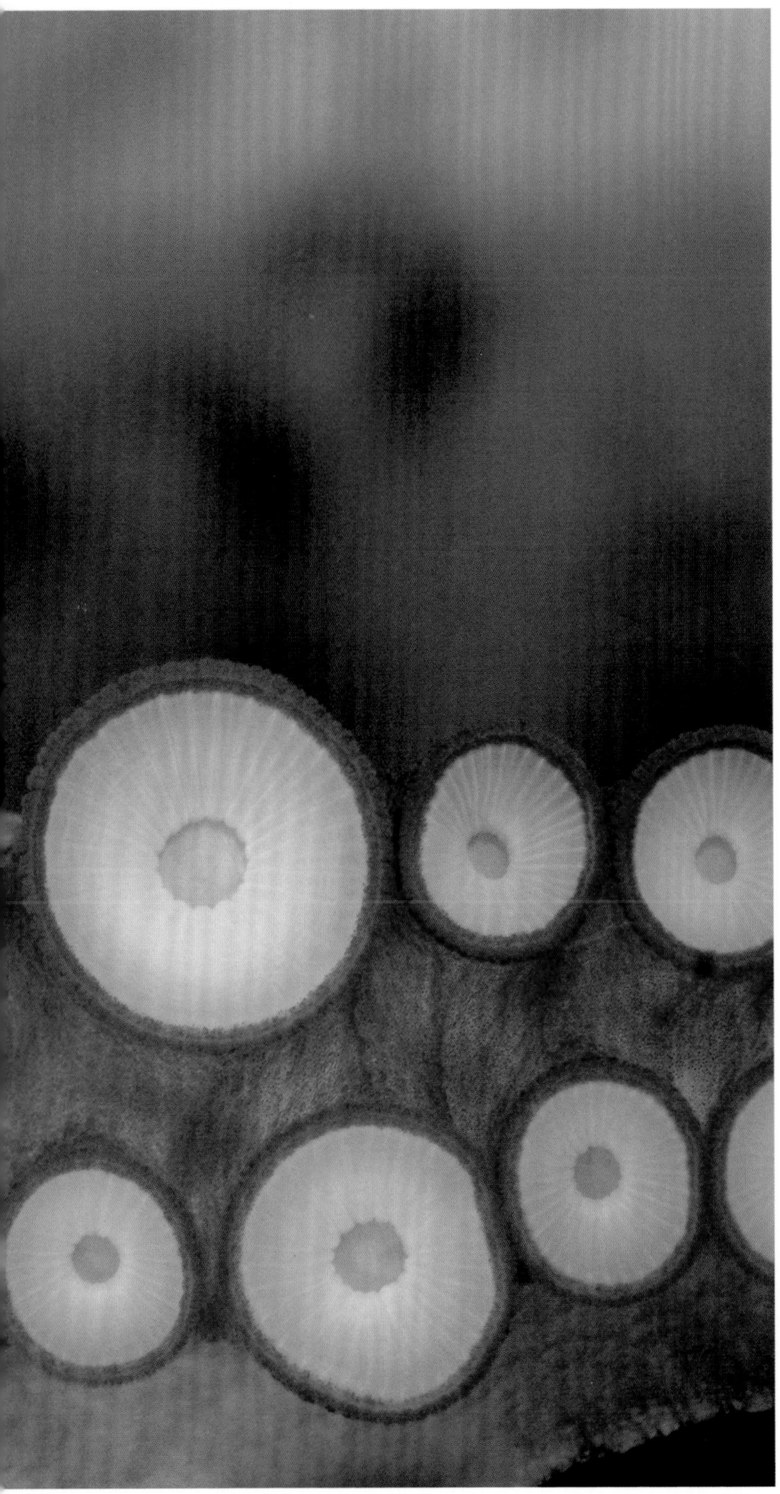

Plate 32. A male common octopus (*Octopus vulgaris*) holding on to the front of an aquarium in Vienna shows the large sucker near the base of one of his arms. Konrad Lorenz Institute. James B. Wood.

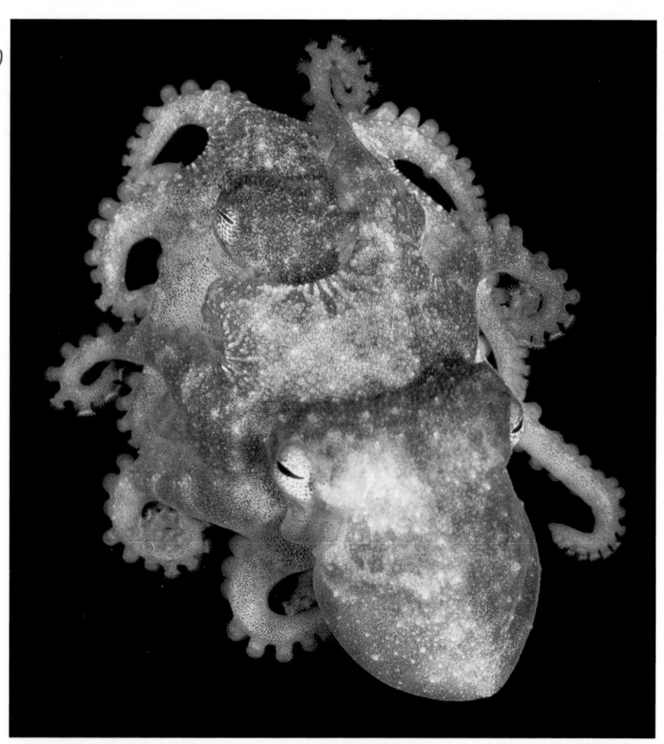

Plate 33. This pair of star-sucker pygmy octopuses (*Octopus wolfi*) is mating in a mounted position. Roy Caldwell.

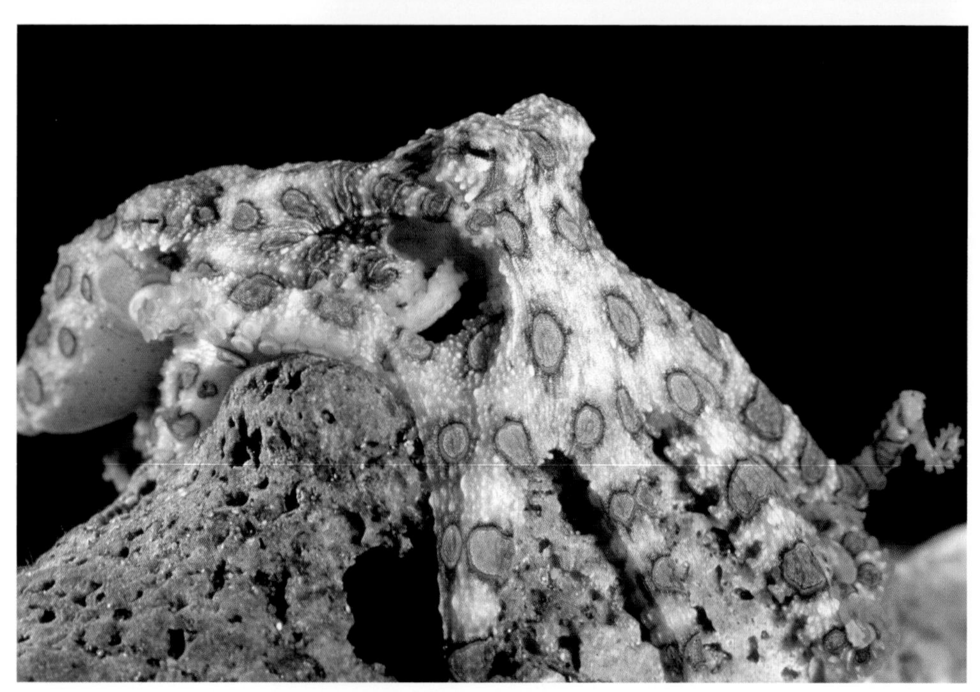

Plate 34. These two blue-ringed octopuses (*Hapalochlaena maculosa*) are mating in a mounted position, so close it's hard to tell they aren't just one animal. Roy Caldwell.

Plate 35. Male Caribbean reef squid (*Sepioteuthis sepioidea*) have visual display contests over status and access to females. The lower male is making a more intense zebra display and will win this contest. Bonaire. James B. Wood.

Plate 36. This little adult pygmy squid (*Idiosepius pygmaeus*) in Phuket, Thailand, is attached to an underwater leaf by her dorsal surface—a good hiding technique. James B. Wood.

Plate 37. The common octopus (*Octopus vulgaris*) does well in captivity. Bermuda. James B. Wood.

Plate 38. The Californian two-spot octopus (*Octopus bimaculoides*) is a good choice for keeping in an aquarium. John Forsythe.

8
Personalities

"Personalities?" We can hear the skeptical reader asking, "Octopuses have personalities?" The simple answer is yes. When a species like the octopus has a big brain and is highly dependent on learning, different members of the species are going to behave differently, especially when faced with challenges related to survival.

Roland got us thinking about octopus personalities in 1987, when he wrote an article about the three species of animals at the Seattle Aquarium that volunteers gave names to: seals, sea otters, and octopuses—two mammal species and one cephalopod species. In the octopus group, there were Lucretia McEvil, who tore up everything she could in her tank, Emily Dickinson, who hid permanently behind the tank backdrop, and Leisure Suit Larry, described as ripe for citation of sexual harassment for excess touching if he'd been human. Volunteers gave these animals names because different individuals were behaviorally very distinct from one another, a characteristic they didn't see in fish or birds but one that reminded them of people.

Ten years ago, not many scientists were saying that there were individual behavioral differences in animals. Science had made a huge step forward in going from anecdotes about one or two animals to data from many. We were systematically looking at the average, making conclusions for populations or species but not for individuals. We said pigeons could home over a familiar range of up to 16 mi. (26 km), that mallard head tosses have a specific angle and differ from pintail head tosses in angle and duration. We might have said that rats can learn a T-maze in an average of XX trials, Y fewer than mice. Our ability to predict what members of a group could do was powerful.

The roots of Western science's attitudes toward animals are in two sources. One is our belief in objectivity—that we can make value-free observations and experiments to get the truth about behavior. Eastern philosophers hold the opposite attitude—that when you observe something, you inevitably change it. Turns out, we are learning that objectivity isn't really possible. Hank Davis and Dianne Balfour challenged the objectivity belief in 1992, collecting a series of accounts by behavioral scientists about the bonds between human scientists and their animal research subjects; Jennifer wrote an article, "Underestimating the Octopus," included in that book. But the belief in objectivity is only slowly waning.

The other source of our attitude toward animals is a holdover from seventeenth-century philosophy, specifically that of French philosopher and scientist René Descartes, who stated that animals are no better than automata (pieces of machinery) and that therefore we may treat them as we like. He separated humans, who have souls and minds, from animals, which are just bodies. Animals were assumed to be simple and their reac-

Personalities?

When I was writing up my observational studies of juvenile common octopuses in Bermuda in the early 1990s, I began to notice behavioral variations among the animals. Some octopuses in the species were very active, and others were seldom out; some liked rocky homes, while others were more likely to be in sand and rubble. One octopus sat at its den entrance for ten minutes before leaving for a hunt; we could always tell by a particular posture when it was ready to go. Others just left quietly when our backs were turned.

The most noticeable behavioral difference was in the octopuses' reactions to us. One team member swore that the long-watched octopus #5 would give a bold visual gesture to the watcher before it left home to go hunting. On the other hand, shy #30 was in one home for four weeks, and despite regular checking, we never saw it leave or return. I'd have believed it never went out, but crab and snail shells appeared regularly in its den midden. From these observations, I was ready to think of octopuses as having personalities.

—Jennifer A. Mather

tions automatic, and lowly invertebrates were even simpler. This attitude made conducting research on animals easier to justify.

From two different perspectives, psychology and ethology, students of animal behavior in the mid-twentieth century continued this tradition of seeing animals as things. Psychology was dominated by behaviorism. The animal was a black box into which we didn't need to peer, with no mind to complicate the learning process. Scientists put stimuli in and responses came out; they could make and predict learning curves and learning laws

Ethics and Invertebrates

Over the past decades, we as researchers have struggled to bring our consideration of the animals we work with into line with the ethical principles we hold. Ethical consideration of animals has followed our general human-centered principle: the more they look like us, the more likely we are to consider their welfare. As a member of the Animal Behavior Society's Animal Care committee reviewing ethics in research, I found myself standing up for invertebrates. In 1987, the Canadian Council for Animal Care (CCAC) guidelines classified octopuses and other invertebrates as tissue, and said researchers could do anything they like to them. I became a member of a CCAC subcommittee that changed this designation a year later. At the Animal Behavior Society workshop on animal care issues in 1989, I again advanced the case of the invertebrates, and became known as an authority on invertebrate awareness and advocate for their care. In 2001, I was asked to write about invertebrate animal suffering in the *Journal of Applied Animal Welfare Science*, and later was contacted as an advisor for ethical standards regarding invertebrates in the UK, Scandinavia, and eventually for common ethical guidelines for the European Common Market countries. Roland and I wrote a review for a special issue of *Diseases of Aquatic Organisms* (2007) on ethics and marine organisms (scientists are just beginning to think about fish welfare), linking reasons for our consideration to different philosophical ethical approaches. But in the United States today, there are still no ethical standards for care of invertebrates.

—Jennifer A. Mather

but not worry about individuals. Personality was just "noise," variation that could be eliminated by getting a large sample size. Ethologists of the time were equally uninterested in individual variation. They wanted to study the species-typical patterns of behavior that happened in the animal's natural environment.

When scientists studied octopuses as simple organisms, they didn't fare well using either of these approaches. For instance, psychologists had the simple belief that with more training, the choice-reward link for the subjects would get stronger. As we know, in octopuses there's no clear choice of prey. In the lab, in a simple choice-rewarded experiment, an octopus spent some of its time checking the unrewarded stimulus and never got to the accepted criterion of eight out of ten correct choices. It stayed curious about what else might be going on and checked the alternate regularly. Cephalopods have so far shown so little stereotyping of action sequences that their behavior doesn't fit the fixed action pattern model of ethology, either.

When and why did scientists abandon pursuit of the average output for the group? For one thing, the ideal of animal brains as stimulus-response relays was fading fast in the second half of the twentieth century. Clearly, learning was not the same in all species, and the idea of constraints on learning was an exciting new one. People began to wonder whether it was a bit limiting to regard mind and consciousness the domain of only humans. We have always looked for exclusive abilities of humans, from language to tool use to consciousness, only to discover that we share these abilities with other animals. If we share 99.5 percent of our genes with chimpanzees, it's not surprising that they share many of our abilities. Donald Griffin's claim (in 1981) that animals must have minds although simpler ones than ours, and so they could have awareness, was scorned for years, in part because it's impossible to prove. But this new approach made a dent in the view of animal as machine.

It became clear that intelligent animals, and some we would class as not very intelligent ones, didn't just give automatic reactions to external situations. Otters play, rats explore mazes before they make their choices, even bees seem to have memorized a map of their neighborhood. Scientists started examining different strategies for getting variable reactions to the same goal, and a whole new area—game theory—grew as part of a new direction of behavioral ecology. Given a difficult but important choice, like aggressive competition, an animal might choose an evolutionarily stable strategy, the same one each time. Or it might vary the strategy depending on the actions of others. In aggressive contests, one animal might choose a hawk strategy (attack first) or a dove strategy (hang back and make peace).

Similar variations in sexual strategies result in mostly male animals of many species responding to sexual selection and having two basic choices. They either dominate because of size and ability and hold territories or defend females, or, being smaller and weaker, hang around as satellites or even "sneakers" that look like females, waiting for a quick chance at reproduction. Giant cuttlefish (*Sepia apama*) use some of these strategies, with smaller males staying near consort pairs and pestering females, while larger males ward them off. Given the chance, females of that species will take the quick mating opportunity with a sneaker male—they like variety in their mates. Octopuses are less obvious about competition, but males have been seen fighting at females' dens. Female octopus sexual strategies center around choice—deciding which and how many males to accept, and when and for how long.

Studying animal strategies has helped us identify variability in behaviors, such as the octopus's food choice. Handy as the average is, it seldom matches the dynamic nature of what's going on in the real world. Behavioral extremes may matter a lot, especially if you have a variable environment, such as the subtidal areas of Alaska where some giant Pacific octopuses live, and selection on the individual level may batter the mean from year to year. Bottom trawling for oysters can drastically change a marine habitat, for example, wiping out several food sources so only generalists like the octopus will survive.

To study individual differences among octopuses, we realized we had to find variation, not minimize it. So we needed to bring octopuses into the lab to give them all the same environment of an aquarium tank and put them in the same simple situations. We settled on three areas: Alerting—opening the tank lid, Threat—touching the octopus with a test-tube brush, and Feeding—dropping a crab into the tank.

Our choice of octopus species to test for personality stemmed from convenience rather than deliberate selection; we worked with what Roland had available at the Seattle Aquarium. We chose the small red octopus rather than the giant Pacific octopus because of size: you can keep about fifty little red octopuses in the same volume of water as one giant Pacific octopus. We would have liked to choose our test species based on which would give us more variability. Every aquarium keeper or researcher who

Animal Temperaments

To measure variability in octopus behavior, being a psychologist I turned to the human personality literature to see what I could adapt to a comparative approach. The work on human personality didn't help me much. Neither Freud's personality stages, which he saw as fitting different sources of sexual gratification, nor Abraham Maslow's levels of self-actualization in finding deeper meaning in life fit well for octopuses. I had more luck when I read the developmental psychology literature on temperament. I still find valuable Alexander Thomas and Stella Chess's 1977 division of behavior into simply described categories: motivation as the *why* of behavior, temperament as the *how*, and abilities the *what*. Personality isn't what you do but how you do it. Developmental theorists agree that temperament is the outline of simple emotional variations, has a strong biological base, is very heritable, and occurs early in the lifespan. Personality is the wider, deeper outcome of one's environmental influences—experiences laid upon this temperament base.

Temperament theorists talked of traits whose amount was partly experience and partly due to your physiology and could place you somewhere on a continuum. The shy versus bold behavior measurement was an example of one dimension where individuals varied: you could be very shy or somewhat shy or a little bold or very bold. Certainly octopus #5 in Bermuda was at one end of the range and octopus #30 at the other end. The temperament approach has turned out to be very useful for studying octopuses. The classification is broad enough to see what is there, yet it doesn't impose outside measures on the animals and allows categories to be sorted out from their behavior.

—Jennifer A. Mather

---

has worked with more than one species knows they are different, and Roland and James have been compiling the trends. Besides being great escape artists, common octopuses are notably active and feisty. Other species, such as the red octopus and the giant Pacific octopus, are less so, though one member of Jennifer's pygmy octopus species would kill another if two were kept in a small space. Still, we needed a large number of octopuses to do these tests, so we chose the red octopus and tested forty-four of them.

Roland conducted the tests for the study. He kept three octopuses at a time, each in its own tank, in the basement of the aquarium, and he tested every second day over two weeks. When finished with each phase, he took the octopuses back to the bay, let them go, and caught another three from a different place. About once a month, Roland mailed Jennifer the numbers to turn into data sets. It took three years to test all those octopuses and record the presence or absence of nineteen likely behaviors when each animal was exposed seven times to three different situations: Threat, Alerting, and Feeding.

Jennifer used the computer to make sense of the numbers. To arrive at personality or temperament dimensions in animals, you record reactions they make to certain situations and you establish consistencies. The test-tube brush touch test, Threat, was a good one: reactions varied widely, from jetting away from the brush to the other side of the tank and inking, to turning toward it, grabbing it, and pulling at it. Reactions to opening the tank, Alerting, and to the crab, Feeding, were less striking, but there were differences. For Feeding, some swam to the crab and engulfed it, others sat in a corner and waited for the crab to wander over, and some waited until night to eat.

Analysis of the nineteen different behaviors was difficult, but the computer eventually put the variations into three different personality dimensions. One was shy-bold: a shy octopus would hide when a diver came anywhere near, and a bold one would stand its ground at the approach of a diver. Another was reactivity: if you touch octopuses with a test-tube brush, the reactive one would jet away, squirting out ink; but touch an easygoing one, and it would just sit there. The third dimension was activity: an active octopus would be like our #5 in Bermuda, always out and going, while a passive one, like #30 in Bermuda, never seemed to come out of its home. There are consequences to these differences in behavior: #5 was killed and partly eaten by a predator during our observation period (see plate 28).

After doing this work tracing personalities in octopuses, we made another claim that octopuses should be considered along with "higher" animals, for whom variation like this is taken for granted. Personality, which is a construct to explain variation among individuals, isn't just something we humans and perhaps monkeys and dogs have, but it is found among all complex animals. There are differences too, of course: sociability is a personality dimension in the group-living mammals, but it's not present for the solitary octopus. But the similarities are striking, and the octopuses add a new scope to the study just by being there.

Bringing the octopus into the study of personalities, the first time for invertebrates, reminds us of how strongly scientific research focuses on a few vertebrates and how little we know about the invertebrates that make up 99 percent of all species on earth (see plate 29). Jennifer coauthored with David Logue a book chapter on personality in invertebrates, which clarifies our lack of such information. Other existing studies on invertebrates have taken differing approaches. David Sinn and Natalie Moltschaniwskyj (2005), for example, continued classic personality research on the little, short-lived southern bobtail squid. Genetics researchers focused on the well-known fruit fly and dug into the physiological and genetic background of simple behaviors like courtship. Behavioral ecologists have studied the evolutionary-linked adaptive strategies that result in our description of shy or bold animals such as spiders, and are gradually coming to see that these behavioral syndromes produce a variety of compromises between staying safe and getting enough food.

Samuel Gosling turned the focus around in 2001, arguing that research on animal personalities can give us a new perspective about human personalities. He argued that through animals, we can learn about the biological basis for human personality reactions and the hormonal and physiological aspects of those reactions. We can learn about the genetic basis: heritability is hard to trace in humans who reproduce after twenty-five years, but is much easier in rats with one year to maturity. Findings can be achieved even quicker for the little bobtail squid, which lays eggs at three months, or in fruit flies with a lifespan of a couple of weeks. He also points out that we can study environmental influences on personality in animals by isolating monkeys or enriching rats' environments in ways that we cannot ethically do with humans.

While octopuses don't fit neatly into this area of research, Sinn et al. (2001) have begun to answer some of the questions about cephalopod personality and temperament. When still a master's student at Portland State University in the late 1990s, Sinn studied temperament in young Californian two-spot octopuses. He deliberately chose this species because it lays big eggs for octopuses and therefore has big young that walk away at hatching and can be more easily reared and tested. By testing at three, six, and nine weeks of age as well as testing ones from different broods, he obtained some interesting comparisons.

He used the same three tests that we did, and found similar personality dimensions in these tiny octopuses (which are about the size of a human little fingernail, though big for an octopus). And because he tested them over time, he could look at other effects that might be related to genetic programming of changes or situations where ecological influences would matter. Most behaviors were the same ones we saw, but change of pupil size, papillae height, and posture seemed more important in these little octopuses. His analysis picked up four factors explaining 53 percent of the variance, one more factor than ours and 9 percent more variation. He chose four names to describe the factors: active engagement, arousal-readiness, aggression, and avoidance-disinterest—which are much like those we gave to the red octopuses.

The second part of the analysis was equally rich in findings. A theme of vertebrate personality development is "continuity plus change." Six weeks' testing time doesn't sound like much of a time span, but this species lives only six months to a year. There was change, significantly, between weeks three and six for all the octopuses. These changes were mostly in arousal-readiness and aggression; maybe wariness and aggression need to change in the early, risky life of octopuses. Still, there were big correlations between scores across time periods, so there was continuity as well. This neat fit with vertebrate developmental patterns is particularly thought-provoking, especially since all the octopuses were raised in tiny, bare isolation chambers. Some internal influence, not a variable environment, had to be making the difference. And who knows how different they might have been if they had been out in the vast ocean—one on a rock face full of algae, one in a rubble-filled yacht basin, and a third near a pebbled beach.

The third interesting result came about because Sinn worked with octopuses from different broods. He collected mature females ready to lay eggs from the wild, so he didn't know paternity. Given the chance, female octopuses will accept several males and store sperm from them. When the eggs are ready to hatch, they move past the sperm storage gland where the female releases sperm to fertilize them on their way to hatching. We know only that the baby octopuses from one brood were at least 50 percent related, having the same mother. Still, the pattern of development of temperament dimension was significantly different for these three broods, with arousal-readiness differing significantly. Some genetic factor was affecting the way these baby octopuses approached their environment, and with a variable environment of the near shore, these inherited differences would probably have been exaggerated. The same kinds of differences exist in newborn garter snakes but within a brood, and this is an interesting parallel because neither species receives parental care. Newborn monkeys are buffered from the environment by their mothers, whereas snakes and octopuses have to make it on their own from birth.

Sinn's work makes us more comfortable drawing parallels between vertebrates and cephalopods about temperament and personality. And his work is comforting in another way: he has picked up our ideas and developed them in his own direction, so the important work of figuring out how and why individual octopuses act differently will continue. But we still wonder about the octopuses we encountered. Did bold #5 in Bermuda have a peaceful childhood, with abundant crabs to eat and no menacing fish? Would its sunny personality have made it vulnerable to the Portuguese fishermen who caught some of our octopuses at low tide? Did shy #30 have a pair of anxious-avoidant parents? And if so, then how did they manage to get together to mate? Maybe his mother's anxiety made her a good caregiver for all those eggs. And what about Leisure Suit Larry? Did a danger-filled childhood wreak havoc on him as it does with men who end up with Antisocial Personality Disorder? Would the octopus Emily Dickinson have been able to tolerate the presence of a male long enough for mating if he'd been able to find her behind the tank backdrop?

9

Intelligence

Most everyone has heard that octopuses are intelligent. But it's difficult to define this intelligence. We know that octopuses learn well—that they change their behavior because of information they get from their environment. Common octopuses have demonstrated this learning ability in numerous situations, such as when they have avoided stinging sea anemones on hermit crab shells, drilled at a different location on snail shells because they were blocked from the spire, and learned to take a crab only when it was associated with a striped visual stimulus. People jokingly refer to octopuses as "smart suckers" with "soft intelligence" and "spineless smarts," but they often wonder just how smart octopuses are. Because intelligence is such a variable thing, made up of different abilities and used in different situations, the question really can't be fully answered.

The book by Marc Bekoff et al. (2002) about animal behavior is interesting because it offers accounts of various animals' abilities by researchers who work with different species—though, alas, nothing on octopuses. Ground squirrels size up potential predators, jumping spiders may use deception, and dogs understand human gestures. These abilities are all different, and each is adaptive in the life of the animal that uses it. Because there's more going on than just simple stimulus-association learning, we like using the term "cognition" for these abilities. Ulric Neisser defined cognition in 1976 as "all the processes by which the sensory input is transformed, reduced, elaborated, stored, recovered, and used," though every animal will have a subset of ways to use these processes that is matched to its ecology. That means intelligence is not only about getting but also using information. Octopuses use information when they construct dens, make a sequence of different appearances to evade predators, and discover different ways for opening a tightly shut clamshell. Much of the octopus's daily survival is based on getting information and using it well. We cannot talk about how smart octopuses are, but we can talk about how they are smart.

We felt we were onto something about intelligence in common octopuses when we charted their foraging areas in the mid 1980s in Bermuda. We knew that these octopuses had a home in which to hide and went out hunting from it, but we were startled by the systematic way they did this. Mostly hunting excursions were short trips of less than an hour, and the route outward was wandering and indirect. One octopus would set out northeast, skimming the rock face but concentrating on the crevices, feeling its way through an algae patch, and doing webover searches through an area full of small rocks. After getting a few file clams, the octopus would stop in a crevice to eat them. On the way back, it would take a direct path. Often, if the animal were more than a couple of yards away, it jetted straight home, making a triangular path. The next time an octopus went out to hunt, it started off in a different direction, and the third time it went in a different direction yet. This pattern revealed that they were central-place foragers, hiding in the middle of a home range and going out into it to find food. This mode results in the animal covering different parts of its home range each day, keeping near shelter but being able to eat.

This behavior tells us something about memory in the octopus. Systematically coming back home from foraging means that octopuses have spatial memory—memory of where home is. We became convinced of this ability when we watched them take detours. One octopus was coming home from a two-hour hunting trip when it got scared off its route by another octopus. Finding itself 9 ft. (3 m) from where it had started, it followed the rock edge along a new route until it got home again, navigating by what it had learned in the past (see plate 30).

This technique is quite common among a variety of animals for getting around. Female wasps called bee-wolves dig a hole in a sandy area, lay an egg in it, and go off foraging for caterpillars that will be the hatching larva's food. In a pioneering 1932 study, Niko Tinbergen (1972) tested the bee-wolves' location memory, or how they found the hole, in a simple way. He made a set of landmarks by placing a group of pinecones around the hole, and the bee-wolf got used to the cones being there. But when he moved the location of the pinecone circle, the insects came back to where the cones were and not to where the hole was. We tried this same test on common octopuses in their natural environment, using a vivid black-and-white cylinder as a landmark. Our bold common octopus #5 was a particularly useful test subject because it didn't mind going out hunting when a snorkeler was lying in the shallows nearby watching it—not every octopus puts

Smart Suckers

I remember clearly when I decided that I needed to regard the octopus as an intelligent animal. I was observing octopus activity in Bermuda in the mid 1980s. I had a lot of help, and we were following two animals through their whole day's activity, from 6 a.m. to 6 p.m. for about four days. I was taking my turn, floating on the surface and watching an octopus that had gone out hunting, returned to its den and had eaten, and was now making casual arm actions that cleaned out the den. Suddenly it took off from the den, traveled a few feet to the sea floor below, and picked up a small rock. It immediately swam back to the den and dropped it, then repeated this action with two more rocks, piling the rocks up in front of the den entrance. Then it entered the den and went to sleep.

While watching the octopus sleep, I had lots of time to think about what it had done. Earlier, when I was a graduate student, I had given choice tests to octopuses for different characteristics of dens, so I knew they liked dens about their volume and with a small entrance. But this case was different. The octopus hadn't chosen a small entrance; it had decided to make a large entrance smaller. To do this, it must have had some idea of what it wanted—known in some way that a pile of rocks would make the den entrance smaller—and then looked out across the sand to see suitable rock candidates and gone out and picked up the right number.

In describing what the octopus had done, no matter how I tried, I found myself needing to include words like wanted, planned, evaluated, chose, and constructed—words that animal behaviorists of the time (and even now) were not likely to use regarding invertebrates. The words moved the animal out of the category of reactive plodder to that of thinking and anticipating being. The observations were, it occurred to me later, also evidence of tool use. In the 1980s, scientists were only reluctantly admitting that some of our monkey relatives might use tools; tool use was "a sign of intelligence" and for years was the hallmark of the cognitive advantage of humans.

It took me a long time to do studies on exploration and play, personality, and problem solving to demonstrate to others that octopuses were intelligent. But from that time on, I knew they were.

—Jennifer A. Mather

up with being observed. When we moved the landmark after three days, the octopus came back to its home and not to the landmark. We suspect this was because the octopus lived in a rocky area full of other clues as to where it lived, not a flat sandy area like the bee-wolf. We found that the red octopus, which hides in beer bottles out on the open sand, does use landmarks for finding home when we tested it in a circular tank in the lab.

This kind of field observation can be used for constructing an interesting lab test of spatial learning. Jean Boal et al. (2000) kept octopuses in a tank that had two small deeper areas at one end, and gave them a few days to explore their new home. Then they drained the water out of the shallow part, leaving two deep refuges, like tide pools. The octopuses remembered the locations of the deeper areas and learned to go to one of them until the water was restored to the rest of the tank, as would happen when the tide comes in (octopuses can survive for a while in air but prefer to stay in water). It's a good memory test, using an ability the octopus has to practice each day.

Octopuses don't hunt in areas they have covered in the past few days. This win-switch foraging strategy is very efficient. If an octopus catches a crab under a particular rock, it shouldn't bother going back to that spot because there won't be another crab there for a while. This behavior in the octopus supports the idea that it has working memory; it remembers not just where things are, like its home, but it also remembers where it has gone for food. This situation has been used in testing animal memory. Rats were given a radial maze that has several arms built out from a central platform, each with food at the end. The rats could go anywhere to get food, but they learned not to revisit arms already plundered.

We made a water version of a radial maze and tried to prove that octopuses had this ability in the lab too, but the octopuses didn't seem to learn to go to still-baited arms that they had not explored before. Later I learned why this experiment didn't work. I tested the small and easily available red octopus, but it turns out that this species may not return to the same home day after day. The test might have worked with a common octopus or a Hawaiian day octopus, both of which do.

This win-switch pattern of foraging areas gives us a clue as to why a lot of learning researchers might have had trouble getting octopuses to make consistent choices of the rewarded rather than the unrewarded stimulus. The octopuses would only choose correctly seven of ten times on a series of trials. Rats made more and more positive choices the longer scientists tested them, and eventually got to a perfect ten out of ten. Octopuses just kept randomly trying the other, unrewarded, choice once in a while. This makes sense from the viewpoint of an octopus that has lived in the sea for months before it was brought into the lab. Food isn't available again in the sea where you just found it, so why try? In a week or so, it might work, since prey animals always need shelter. So octopuses keep trying the alternative choice some of the time, because that approach works out in the ocean.

Octopuses also use learning to direct their actions. A simple type of learning is habituation, when an animal stops responding to some repeated and uninteresting situation. Humans use habituation all the time—no longer noticing the hum of the air conditioner or hearing the piano practicing coming from next door unless it's particularly off key. Octopuses may well use it too. Many animals, including the simplest worms and sea anemones, will habituate to repeated stimuli in their environment. But we found that habituation had been little studied in octopuses, so we, and Michi Kuba and colleagues (2006), took it on.

Kuba continued to study learning in common octopuses along the same lines that we had studied earlier in giant Pacific octopuses, showing their balance among exploration, or getting information out of the environment; habituation, or losing interest; and play, or manipulating the environment or the social interface without an obvious immediate purpose. It's seems logical that octopuses would explore, since they have to get information to survive in their variable environment. In the studies, the octopuses started off a sequence of trials by grabbing a floating pill bottle or three-dimensional plastic block we gave them. Then they pulled it to the mouth and held it with all the suckers on the arm bases. As you would expect, after a while the octopus held it less tightly, eventually lost interest, and stopped exploring it as the trial wore on.

When we gave the octopus a test object such as a jar (see plate 31), it would explore it until it got tired of the object. When Kuba used as his stimulus object a plastic crab model outside the tank, the octopus quickly learned it had seen it before and couldn't do anything with it, and lost interest. But when the octopus could manipulate the object, results for habituation among tests wasn't consistent—there was octopus variability. Part of the reason that habituation wasn't consistent is that we began to see play, another action only higher vertebrates are supposed to do. Animals with a lot of manipulative ability are able to move in viewpoint from

"What does this object do?" to "What can I do with this object?" Instead of losing interest, they switch approach. The octopus Roland first saw doing this play behavior blew jets of water from her funnel to send the pill bottle to the other end of the aquarium where the water intake flow sent it back to her, and she repeated this action twenty times. He didn't see a consistent decline in response, but instead saw a change in the type of response. The playlike behavior peaked at the third or fourth day, then the octopus seemed played out and decreased the contact time.

Can we say that an octopus is intelligent by seeing it displaying play-like behavior? Playing with objects isn't about learning, in a simple sense. An octopus will learn about an object or its own ability when it plays, but it also will do so when it just explores an object. So far, only animals that play are ones we class as intelligent, though this may be another case of anthropocentric humans needing to look further. Still, this whole set of behaviors makes us see that the octopus is intelligent. If we want to understand how animals gather and use information, we have to look at variable exploratory behavior. An octopus's going out to see what's there, feeling around to understand what things are made of, then manipulating an object to see what it can do with it is interesting behavior for study.

Another important question about intelligence and the role of learning in animals' lives is how an animal changes across time and with development. This question has barely been looked at in octopuses. But cuttlefish have turned out to be a good subject, because the hatchlings are fairly large for a cephalopod and settle onto the bottom quite soon after birth. They are unlike octopus paralarvae, which are usually tiny and float off into the plankton for weeks, and even octopus benthic juveniles are very small.

It is easier to evaluate prey capture in newly hatched cuttlefish, which are very picky about their food. John Messenger showed in 1977 that they would only take tiny mysid crustaceans or models that looked like them. When mysids were put in a tiny glass test tube and the little cuttlefish struck at them with their tentacles, they couldn't learn to stop. At the beginning of life, these animals had a narrow, fixed search image. They did not need to learn what food was appropriate to catch but they couldn't learn about new food either. As several weeks passed, the cuttlefish began to learn to try for different prey species—their search image expanded. They also learned better and better not to try to catch the mysids in the test tube. The cuttlefish were able to accomplish short-term learning, over about five minutes, after one week of life, and they got the ability to learn long term, over one hour, by one month of age.

We believe that learning is an ability that's found in the brain, and visual learning in cephalopods is related to the brain's vertical lobe. If the lobe is surgically removed, an octopus can no longer learn. Messenger also discovered that newly hatched cuttlefish had only a tiny part of their vertical lobe. When several weeks passed and the cuttlefish could learn, anatomy studies showed a big development of the vertical lobe both in size and connections to other brain areas. Gradually a learning-dependant animal learns to eat more variety of what's around it as it has time to find it. As well, both an enriched environment and the presence of particular prey species would change the learning program in those tiny animals, so later they accepted those prey species more readily. Environment had a tuning role for learning what to eat, right from the start.

Recent research has shown some early learning in cuttlefish. Ludovic Dickel et al. in 2006 found that cuttlefish aren't always tightly bound to a narrow recognition of mysids as prey. In their first week of life, tiny cuttlefish still in their translucent egg capsule could see general shapes. If they were shown small crabs, they later changed their prey choice and also accepted crabs. Such early learning is much like that of humans, and could be called imprinting.

This learning behavior has a clear parallel with that in young mammals, including humans. At birth, humans have a set of reflexes—automatic actions that don't get changed by learning and help us to survive. We can grasp with surprising strength, which helped our monkey ancestors not to drop from mother's chest to the forest floor. We also have a sucking reflex that helps us benefit from the easily available and nutritionally perfect milk of mother's breast. In fact, newborn human (and many other mammal) babies can do something that adults lose the ability to do: suck liquid and breathe at the same time. They roll up their tongue into a flexible tube near the roof of the mouth to send milk down one way and receive air down another. These reflexes and more are gradually lost by six months of age as the baby also begins to learn to tune feeding and response to caregivers and to what is around them. Actually, they can even tell their own mother's milk from anyone else's by one week of age (Charles Maurer and Daphne Maurer 1988). The beginning stage with automatic responses and no learning is followed by a gradual replacement with learned associations. The shift is the same one experienced by cuttlefish, the best developmental program for survival of an inexperienced newborn that will later become a learning specialist.

This study is an example of a larger area of research known as constraints on learning. Scientists have found that, generally, animals get information from their environment but that heredity controls what and when they learn. Small birds learn who their mother is, and human babies learn the visual cliff—something that looks far away, and down, and shouldn't be walked to—and that each animal has specific times of development. We can imagine that octopuses have these time-sensitive periods, such as when they settle out from the plankton. Settlement is likely a sensitive period of fast learning for young octopuses. Settling paralarvae of small-egged octopuses would have to learn where to hide, how to color match the background (which in tiny cuttlefish is affected by experience), what kind of food is edible, and even how to walk and catch prey with their newly elongated arms.

Regarding the program of learning, an octopus or any animal that depends heavily on learning will have its learning guided by heredity. After all, learn wrong and you may be dead. We have gradually learned about these constraints on learning. Bees learn about flowers' color, smell, and shape so they can find them again, but they learn some colors better than others and can't learn some combinations of these cues. Gerbils will work long and hard for sunflower seeds as a reward, and octopuses consistently go for crabs. Early octopus learning researchers found that they couldn't treat a crab as a negative stimulus, when the octopus got shocked for touching it; the octopus just couldn't learn not to try to catch crabs.

Some other constraints on learning that octopuses have might tell us about their anatomy and the way they process information. Wells found that common octopuses can learn by touch and can tell a smooth cylinder from a grooved one or a sphere from a cube. They had much more trouble, though, telling a cube with smoothed-off corners from a sphere, or vertical grooves versus horizontal grooves. They couldn't learn to distinguish a heavy cylinder from a lighter one with the same surface texture. Maybe the common octopus could not use information about the amount of sucker bending to send to the brain and calculate what an object's shape would be, or calculate how much the arm bent to figure out weight. Octopuses have a lot of local control of arm movement: there are chains of ganglia down the arm and even sucker ganglia to control their individual actions. If local information was processed as reflexes in these ganglia, most touch and position information might not go to the brain and then couldn't be used in associative learning.

While that local control may be true for texture information, the studies on hole drilling for octopuses suggest that they have to use the arm postures and place information for other judgments. We know that octopuses drill into snail and clam shells at particular places, and that these are the right places for the toxin they are going to inject to have the most effect on the prey's muscle. A few years ago, the Seattle Aquarium had an octopus that was so tiny it had never seen a shell it could drill. The first few times, it drilled haphazardly in any area of the shell, quite far from the body or muscle of the clam that it needed to attack. Within a few drills, it changed location to aim over the center of the valve, which was over the heart, and it did this for the rest of its time in captivity. This early learning study needs to be repeated with lots of tiny octopuses, but it's difficult to find them. This learning must be related to position sense, since this penetration of the clam's hard shell is done while the shell is held under the arm web and is out of the octopus's sight.

Detour experiments also have helped us discover whether an octopus can remember where to go. In early studies, octopuses saw a crab through a glass viewing area, then had to move aside, go down a corridor, and turn a corner to actually catch the crab. They could only do this, and with difficulty, if they moved down the corridor keeping in touch with the glass wall and consistently looking toward the crab with one eye. After the more than forty years since the studies, we've learned that common octopuses are usually monocular when looking at possible prey. When an octopus learns about a situation by getting information with one eye, it stores the data in one side of the brain, and by the next day, has transferred the information to both sides of its brain. So when crawling down the corridor, the octopus would only have half of its brain to find its way, following along the wall with one eye and one set of arms.

There's another constraint on learning that could have led to these failures to solve the detour problem. The octopuses had to learn that a crab seen through glass was really there, despite the touch information that said the crab was visible but not accessible. Maybe octopuses don't understand what glass is. This must be an unusual piece of information for them, because things visible in the ocean can just be reached out for and touched. Lots of animals don't understand the properties of glass. Birds fly into glass windows, six-month-old infants touch a glass surface over the visual cliff but don't believe their touch information when their eyes tells them differently. This difficulty with glass could account for problems with getting common octopuses to learn to get a crab out of a glass jar by unscrewing the lid. They can do it, but they don't decrease the amount of time they spend playing around with the lid before they unscrew it. In other words, they didn't learn. Could it be that glass just doesn't make sense to an octopus and so it never learns to use its arms any better with it? Perhaps arm use in octopuses isn't as accessible to learning as it is in vertebrates because of the huge amount of local control of arm movement. Maybe only a little information about arm action trickles up the ventral nerve cords to the brain. The octopus definitely has limitations but we're not yet sure why.

Roland started us asking this question in a new way, one that reveals a lot about how science does and doesn't work. He studied a giant Pacific octopus in the Seattle Aquarium that eventually decreased the amount of time to remove the lid of a glass jar in order to eat a piece of herring inside, and so this animal did learn. In this demonstration, Roland didn't carefully wash his hands, and so chemical cues from the smelly herring were all over the outside of the jar. When we tried this more formally with a crab inside the jar, but smeared herring outside, octopuses steadily decreased the time to open the jar lid, a clear demonstration of learning. We know that when the octopus tries to take a lid off a jar, the jar is hidden inside the arm web and out of sight. Maybe the octopus can't remember what it has seen without a little reminder of chemical cues to keep it working, and information from more than one sense makes a useful combination for the octopuses and for us. After all, we look up the apple pie recipe in a cookbook, reinforced with a great picture, and watch it as we build it. But it's the marvelous smell that reminds us it's time to take it out of the oven and eat it.

While there are limitations on use of information, octopuses can learn a lot about visual and tactile stimuli. In this case, it makes sense to talk about cognition as use of information. Norman Sutherland (1960) set out to test what cues a common octopus used to learn about a visual shape. He hoped to find a simple system that narrowly analyzed incoming visual cues. First he tested octopuses on learning to discriminate vertical and horizontal bars, which they could do easily. They were less good at telling oblique bars from one another but so are mammals. They could tell mirror-image (reversed) figures from one another. Then he tried them on something harder, telling the difference in ratio of edge to area, most simply seen in discriminating V versus W. They had no problem, and bees can do this too. Octopuses could tell if the same figure was rotated 90 degrees, though this was more difficult for them.

William Muntz thought (1999) that this simplistic approach to octopus vision wasn't working. He decided that octopuses were not just using one simple visual dimension to tell the two shapes apart. He made up two shapes that weren't different in any of the ways that Sutherland said were important for octopuses. When the octopuses discriminated this pair of shapes, he concluded we couldn't use a simple model of shape information processing. Octopuses didn't have a simple and automatic shape analyzer tucked away in the brain. They were, instead, learning what to learn each time they were given a new pair of shapes.

Discussion of what an octopus uses to sort out visual shapes leads to another fascinating aspect of learning, simple formation of a concept or idea. When octopuses were given two useful cues about a shape—brightness and orientation—at the same time, they learned faster than when there was only one cue. Later, testing on only one cue or the other showed that twenty-two octopuses had used brightness and six had used orientation. Children learn to do the same sorting of useful from useless cues. When octopuses were trained on orientation as a cue for a long time and then switched to shape, they took longer to learn it. This switching off, using an unrewarded part of a stimulus, is something humans learn too. Training with a particular comparison helps the octopuses make a difficult comparison too. When they were given a fine discrimination that they couldn't learn, they could master it if they were given more obvious differences and then finer and finer ones. Before we get too impressed, we should mention that bees also can do this. But this skill is not a trivial one, and not what we think of as simple learning. Rather it is attention or selection, learning what to learn and how to ignore other available information that turns out to be trivial.

Another situation—when looking at mirrors—may reveal something interesting about the octopus's use of information. When we look in a mirror, we know that the image we see is ourself. But what does an animal see when it looks in the mirror? Most species tested so far treat the mirror image as another animal, and aggressive males often posture and make gestures at it. James found cuttlefish doing this when they came by the window of the Aquatron tank at Dalhousie University in Nova Scotia. He said it was a good opportunity to get pictures of the aggressive zebra display. Most monkeys do the same, but Gordon Gallup (2002) has used what he calls the spot test to look for self-recognition. Put a colored dot on a small child's or an animal's face without them knowing. If they look in the mirror and touch the spot on their own body, they know the mirror image is a representation of self. Only the primates pass this test. Even young children don't understand the spot is on them; they need to be two years of age before they touch the spot. Recent work suggests that elephants, with really big mirrors, and porpoises, arguably very smart though marine mammals without hands to use to explore, may pass the spot test too.

It's interesting to wonder whether octopuses would pass the spot test, which isn't a test of vision but a test of self-recognition, and it could be a test for consciousness. With its good vision, an octopus can clearly see the image of the octopus in the mirror. But whether it would know it is itself is hard to tell. Since an octopus does not normally sees itself or guide its arm actions by vision, what kind of information processing would it need to recognize the sight of itself? In addition, octopuses are solitary for much of their lifespan, and those we watched in Bermuda were indifferent to the sight of others of their species. They didn't seem to have a concept of their own species, and even would eat smaller ones. The fact that octopuses will be cannibals if they can, suggests that octopuses don't have an octopus-recognition template in the brain, and this would argue for them not knowing that the mirror image is themselves. But it's an intriguing test not only of dealing with information but also what kind of information matters to the animal. We found that octopuses seeing themselves in a mirror do know that there's something interesting there, but they don't seem to know that it's "me."

While we know a lot of what an octopus can't do, by finding limits to concepts like learning types, concept formation, and self-awareness, we learn about an animal's abilities. Logically, many animals store information for use in quite narrow situations, such as prey choice or mate recognition. Bees use color information in finding flowers but not in other situations, and they use navigation just to get to flowers, no other time. Bees have pretty small brains. Caldwell, who has studied stomatopod crustaceans extensively, praises their intelligence. Others comment that members of this group, such as crabs, seem to be able to learn cues and use them only in the most apt situations. This selectivity of learning is domain specificity.

Like vertebrates, octopuses gain and use learned information much more flexibly, applying it to a new situation. In particular, consider the octopus's use of jetting. They start out using the jet automatically for respiration and waste removal and as a means of fast transport. But soon they put jetting to use in building homes, cleaning out sand and mud, and uncovering crabs hiding in the sand. It's amusing that common octopuses jet to repel scavenging fish that take crab pieces from their middens, and it's even more amusing when they use the jet on divers like ourselves when we check the shells they have thrown out. The giant Pacific octopus in our lab used the jet to do the equivalent of bouncing a ball, when she played with the floating pill bottle. Despite limitations, variability is still the hallmark of octopuses' behavior. That's why octopuses are such fun to watch and why there's so much more to learn about them.

10

Sex at Last

Mating and reproduction in octopuses takes place at the end of their lives, which is the case for most cephalopods. The parents provide no care for the young once the eggs hatch. The males' life work is completed once they mate, and they usually die shortly after. Females die just after their eggs hatch. So for octopuses, mating signifies the beginning of the end.

After octopuses mate and reproduce, males go into a state of senescence—a state comparable to human dementia, in which they don't behave normally. Because octopuses and other cephalopods have short lives by human standards, O'Dor suggested in 1998 that they could borrow the motto of the Hell's Angels biker group: "Live fast and die young."

Aquarium visitors frequently ask, "Why does the octopus die so young? Look at how big it is, only three years old and already senescent." Even the largest octopus species has a life span of three to four years at most, and for the smallest species it's six months or less. Some deep-water octopuses may live longer, but everything is slower in the cold depths. Octopuses don't really die young; they die after a full life (unless they get eaten). Their complete life spans just happen to be a lot shorter than ours. It is rather a shame, though, that an intelligent animal like an octopus dies after one or two years. We wonder what they could do with their intelligence if they lived ten or twenty years!

Before octopuses can mate and reproduce, they have to find another octopus to do it with. Attraction by sight, chemical attractants, and visits to previously occupied dens are several ways a male and a female octopus can get together.

The story of Ursula the Octopus provides clues as to how octopuses might find each other and made us believe in the use of chemical attractants by octopuses. When we released this captive-reared giant Pacific octopus from the Seattle Aquarium to allow her to mate in the wild, we made observations on her survival and behavior. There are few such long-term observations of octopuses in the wild, partly because scuba divers are unable to stay underwater and make observations over long periods, especially in cold water.

A scuba diver had donated Ursula to the aquarium as a very small, about 2-oz. (50-g) female giant Pacific octopus on May 3, 1997. When she got larger, we placed her out on exhibit. She was an aggressive animal, so we named Ursula after the evil sea witch in the Disney film *The Little Mermaid.* We released her into Elliott Bay in front of the aquarium on March 12, 1999, because she was getting too large for the tank. Although her appetite was still normal, she was approaching reproductive size, 45 lb. (20 kg), and we wanted to give her a chance to mate. She had lived in the aquarium for twenty-two months.

Her release became a media circus. In addition to being watched by some 200 aquarium patrons, Ursula's release was covered by six television news crews, and we gave five radio interviews about her. Her release was featured in three local newspapers and on television's Discovery Channel.

To encourage her to crawl out of her tank, we smeared herring on the lip of the tank and down one side. She stretched a couple of arms up and followed the herring smear with her suckers, leading her to crawl out of her tank into a barrel of water on a cart, amid the screams of about a hundred children who were watching. We then wheeled her to a ramp at the water's edge in front of the aquarium and tipped the barrel so she could crawl into the water.

Aquarium divers with underwater video cameras filmed her release and her first forty minutes of freedom. She jetted about 30 ft. (10 m) to the pilings supporting a portion of the aquarium, perhaps to get out of the light, and stretched out her arms to keep contact with the pilings as she slowly drifted to the bottom 45 ft. (15 m) below. She maintained her bright red color as she drifted down. When she contacted the bottom, she changed to a camouflaging mottled pattern with brown and white colors that matched the background. She came to rest under the end of a suspended log, and was still there when the divers ran out of film and air.

In the succeeding weeks, Ursula didn't go far. Two weeks after the release, a scuba diver found her in a den, which was excavated under a partially buried log at the base of a rock pile under a finger pier at the aquarium, 90 ft. (27 m) deep. This den had previously been occupied by other octopuses. Her den midden at that time was littered with the dismembered and empty shell remains of several kelp crabs. The den was cleaned out and the crab shells were new, so we knew she was eating well.

Two weeks later, she was still there. And near her was a larger octopus in a new den in a large piece of discarded PVC pipe approximately 90 ft. (27 m) away. Before Ursula's release, scuba divers in the area had not seen any octopuses. Since this larger octopus wasn't in the area before, we assumed he was a male and that he was attracted to Ursula. We confirmed his sex on a later dive. Two weeks later, another male appeared in the area in a different den, so it looked like Ursula was attracting suitors. These observations suggest that a mature female octopus can attract mates by releasing chemical cues.

As we followed her progress, we saw no matings, but it seems likely that they occurred. Although her first den would have seemed to be a good place for Ursula to lay eggs, she moved into four other ones during the summer of 1999 before we lost track of her. When we last saw her, there had been no new midden material for fifty-nine days. We thought she was ready to lay eggs, since she had stopped eating and had blocked the entrance to her den with empty clamshells, a typical behavior of brooding females. Unfortunately, at about that time, a pile-driving barge moved next to the aquarium to put new treated support pilings into the bottom of the bay, which disturbed the sediment, created noise in the area, and probably put new chemicals in the water. This commotion probably scared Ursula away, since we didn't see her any more after 124 days from her release.

Once a male and female octopus have found each other, how do they mate? Aquarium visitors see all those arms and wonder, what do they do with them? And how can you tell the difference between a male and a female octopus by looking at them? More important, how do octopuses tell the difference between a male and a female?

There is a broad range of sizes at which male and female octopuses mature in most species, and many octopuses are able to mate over several weeks of their life span. So a large female can mate with a small male, and vice versa. There are also broad differences in an octopus's size at maturity within a single species across its geographic range, based on age, feeding success, and water temperature. Mature female red octopuses in California are as small as 1 oz. (28 g) in weight, while in the cold Puget Sound, egg-laying female red octopuses are frequently ten times larger. Mature spermatophore packets have been found in young male common octopuses that are as small as 7 oz. (200 g), even though adult male common octopuses can get as big as 55 lb. (25 kg). Octopuses that live in water at the colder end of their range, like the species in the deep, tend to live longer and grow larger.

Mature octopus males of many species frequently have a few greatly enlarged suckers on the underside of their arms fairly close to the mouth (see plate 32). Scientists have made several speculations about the purpose of these enlarged suckers. One idea is that they are used as a sexual attractant during courtship. The male octopus may exhibit these large suckers to females as an indicator both that he is a male and that he is of good quality for a mate: he is big and therefore able to find lots of food. The other possible use of these large suckers might be to maintain proper position during mating. During one mating posture, a male wraps his arms around a female, and the large suckers may aid in holding this position.

The external structure that distinguishes males from females is the males' modified arm used in mating. Octopuses have eight arms. The specialized third right arm of an adult male, a hectocotylus, is usually shorter than its third left arm. It has a groove running down the webbing of the arm, along which spermatophores pass to a female during mating. This arm also has an organ—the ligula—at the end, which transfers the spermatophore into the female's oviduct during mating. It might also be used to remove sperm from the female from previous matings with different males. We don't know how the ligula finds the oviduct in the female, but the female may release chemicals that help this process.

In a parallel with vertebrates, the ligula of at least one species of octopus has erectile tissues, remarkably similar to those of human sex organs—the first finding of erectile sexual tissues in an invertebrate. Researchers have noted that the ligula of a two-spot octopus was much larger than normal after an unsuccessful mating. They then looked at tissues of preserved ligulae in museum specimens. By sectioning them, they found that the ligulae contain vascularized internal cavities surrounded by walls filled with collagen fibers, all of which are also enclosed by collagen, like in mammal penises. An expandable ligula in a male octopus might be present for several reasons. The male octopus that had evolved a way to make his ligula bigger could transfer more spermatophores, or larger spermatophores containing more sperm, during mating. It may have also evolved to allow the female to recognize a male ready to mate. Seeing the ligula, a female may be gauging quality—deciding that a big ligula means a big and therefore high-quality male—and choosing a mate by the size of his expanded organ.

Having an expandable, erectile ligula may also help an octopus protect it. The ligula of the giant Pacific octopus is uncolored because its skin doesn't contain chromatophores. And so it may stand out when the octopus camouflages to match its surroundings, though most male octopuses keep the ligula curled up at the end of their hectocotylized arm. When Jennifer and O'Dor were studying octopuses in Bermuda in 1991, they had trouble assigning sex to the animals. Females were easy, but males kept their definitive ligula tucked under the arm web, out of sight. If the ligula of this species shrinks when not in use, it would also be less visible.

There is a remarkable story about the naming of the hectocotylus of the male argonaut. The female argonaut constructs a cornucopialike shell with her first two large arms and the webbing between them. She lives in this shell, lays her eggs on it, and guards the eggs until they hatch. The shell is much prized by shell collectors when it is found washed up on a tropical shore after the eggs hatch and the female dies. Incidentally, such washed-up shells were the first evidence of this species: Aristotle knew about the shells but never saw the octopus that made them.

Although the shells and the female argonaut had been known for millennia, males were not recorded until the nineteenth century. Some strange wormlike crawling things had been found on the females' bodies, and the pioneering French naturalist Georges Cuvier (1829) thought they were a parasitic worm. He described the worm, complete with a hundred suckers, as a species new to science. He named it *Hectocotylus,* the worm of a hundred suckers. As scientists of the time learned more about argonauts, some doubted the validity of this description.

One doubter, Albert von Kölliker, published an article in 1845 describing the wormlike thing as the long-lost male argonaut. This was closer to the truth, but the form was actually just the broken-off hectocotylized arm of the male. Because the arms of octopuses have nerves and ganglia that give them much independent movement, a separated arm can wriggle for hours as if alive. The mystery was eventually resolved when Heinrich Müller (1852) published a paper on the true identity of the male argonaut. Males are tiny compared to the females, about one-tenth their size. The male's third left arm has a sac that contains a hectocotylized arm about ten times as long as his body, about 5 in. (13 cm), and half of it is the ligula.

Today, more than 150 years later, we still know little about the reproduction of the argonaut except that the male's hectocotylized arm matters. We know that the hectocotylized arm breaks off and ends up in the oviduct of the female but we still don't understand the process. Argonauts are quite common, and their cast-up shells litter the shores in some areas. But they are extraordinarily difficult to keep in captivity, possibly because of stress experienced during collection or transport. Since the open ocean is a big place, we speculate that males find prospective mates by a chemical attractant given off by the females.

In a mature male octopus preparing for reproduction, billions of sperm are produced in the male's paired testes. In dissection, these organs are white, as they are in other animals such as other invertebrates and fish, so they can be easily found. There is only one duct leading from the testes, the vas deferens. It goes to the spermatophoric complex, where sperm are packaged into spermatophores, and the packets are stored parallel in a structure known as Needham's sac. Spermatophores vary greatly in size by species: giant Pacific octopuses may have only ten large spermatophores ready at any one time, but the common octopus may have 200 much smaller ones. The giant Pacific octopus spermatophore contains up to seven billion sperm and is 3 ft. (1 m) long. A worker at another major public aquarium once called to ask about a possible parasite in a giant Pacific octopus they were keeping. He described it as a clear roundworm a yard long. This was a spermatophore.

An octopus spermatophore is a complicated structure. Thaddeus Mann published a book in 1984 on invertebrate spermatophores, with a lengthy chapter on those of cephalopods. Spermatophores have a clear tunic and are filled with a viscous fluid, a rope of sperm at one end, and a complicated ejaculatory apparatus at the other. When the male places a spermatophore at the entrance of a female's oviduct, the ejaculatory apparatus everts and forces the sperm rope to the distal end, normally into the female's oviduct. In some species, the spermatophore forms a bladder the shape of a hen's egg and about that size before it bursts and releases the sperm. The sperm attach themselves to the walls of a female's special gland, the spermatheca, where they can remain viable for several months until they are used to fertilize eggs laid by the female.

The release and transfer of the spermatophores takes place in an Arch and Pump action. The spermatophore is ejected by the male's ligula into his funnel and he curves his funnel under the web and up to the base of the third arm. Then, in a series of peristaltic pumps, he slowly transfers the spermatophore into the groove of the arm. This 3-ft. (1-m) structure moves down the ciliated groove along the arm. It takes at least an hour for a giant

Octopus Love

Olive the Octopus was first spotted during the annual octopus census I organized to count giant Pacific octopuses in the Puget Sound region. As a biologist working at the Seattle Aquarium, I know that octopuses are immensely popular with the public. As early as 1875, naturalist Henry Lee said that an aquarium without an octopus is like a plum pudding without plums. A recent survey of aquariums in the United States revealed that twenty-four large aquariums display giant Pacific octopuses, the octopus species most displayed.

The annual Octopus Week celebration, of which the survey is part, is evidence of the popularity of octopuses with the public and also features reproduction. I organize this event near Valentine's Day, with an Octopus Blind Date on that romantic holiday, in which male and female giant Pacific octopuses are brought together for possible "octopus love" (which happens on average one out of three times). This part of Octopus Week is a great draw. Other events include public talks about octopuses, arts and crafts activities for children, octopus releases back into the wild for reproductive opportunities, and octopus dissections. The last part of the week is the annual census.

I am often asked how many octopuses there are in Puget Sound, including brooding females, and how the population has been doing over the years. It was to answer these questions that I organized the annual divers' survey of the giant Pacific octopuses in Puget Sound. I wondered whether reproduction was keeping up with mortality. I hoped to establish a baseline of how many octopuses are in the area and establish whether the population is healthy and stable or sound an alarm if there's a decline.

To gather the octopus census information, I ask the divers to report the location of the dive site, where the octopus was in relation to shore landmarks, the depth of the den, the time of day, a description of the den, and an estimate of the octopus's size, including some measurement of the octopus, such as the width of the largest sucker visible. I use the size estimate to distinguish among sightings of the giant Pacific octopus and the much smaller red octopus. Most sightings come from the Puget Sound areas of Hood Canal, the Tacoma Narrows, and Admiralty Inlet. On average, 162 divers per year have reported seventy-two octopuses seen in Puget Sound over a three-day period. Other than fishery statistics, this census may be the only annual survey of octopuses.

—Roland C. Anderson

Pacific octopus to deliver one spermatophore completely to the female. Once the octopuses are joined, during mating the heartbeat of mating males skips a beat at the passing of the spermatophore, and the females' respiration rate goes up.

Before actually mating, animals frequently perform courtship. During courtship, animals establish whether a partner is the correct species, sex, and maturity to breed with. Cross-species sex does occur, but it is futile and a waste of genes. Hybrids resulting from such matings are often stillborn or weak, and at best they are sterile. Courtship helps females determine mate quality, and it may also help bring the partners into readiness for the next stage, actual mating. Few examples of courtship have been recorded in octopuses, because they are difficult to study in the wild.

In 2008, Crissy Huffard and her colleagues documented the amazing sex habits of a small octopus, *Abdopus aculeatus*, in Indonesia. They observed that each male would try to pick a female and guard her closely from other males. They saw smaller "sneaker" males, which put on the colors and behaviors of females and tried to sneak pass the guardian male. The males picked the largest females they could find; larger females have more eggs in which to pass on the males' genes and they can better survive the egg-guarding process. These observations in the wild are valuable, since we haven't seen these behaviors in the lab.

Once courtship is finished, octopuses proceed to actual mating. The method by which octopuses mate has been known for millennia. Aristotle, in the fourth century BCE, stated, "The male octopus has sort of a penis on one of his tentacles . . . which it admits into the nostril of the female." While inaccurate in details (the mantle cavity is not a nostril), his observation was remarkable for his time and essentially correct. Despite a long history of observations, octopus mating behavior has been accurately reported in relatively few species and from fewer yet in the wild.

During mating, octopuses may employ either of two general positions, either the male on top of the female (mounted mating) (see plate 33) or side by side (distance mating). During the side-by-side position, the hectocotylized arm is extended some distance to reach the female. A male may stay outside a female's den, flashing skin displays and watching her behavior to assess her readiness and willingness to mate. He can just extend the mating arm into the den and into the female's mantle cavity to accomplish spermatophore transfer. There is some debate in the literature about whether males grab the female closely or extend the arm from a distance, and this second position for mating may allow a male to court a female while both stay protected inside their dens. A pair of octopuses has been observed in adjacent tanks in captivity doing this, the male's arm extended up into the air, over the tank edge, and into the female's mantle cavity.

Roger Hanlon and John Messenger (1996) classified a number of octopus species according to whether they used mounted or distance mating, but some species used both methods. The Caribbean pygmy octopus used mounted mating in the laboratory and distance mating in a well-spaced seminatural situation, so mating positions may be different in one species in different environments.

While there are approximately 100 octopod species in the genus (and 300 in the order), mating behaviors have been only been reported in a few. The lack of mating observations even applies to the giant Pacific octopus, the octopus most kept by public aquariums. There are two brief descriptions of matings by giant Pacific octopuses, and both were in captivity, so these observations may not be good descriptions because they could be influenced by unnatural conditions of crowding and sudden contact.

The different conditions—wild versus captive, distance versus mounted—and limited observation times for these matings make a thorough description of giant Pacific octopus mating difficult to develop. But there are a few consistent mating features. Males were darker than females, generally a dark reddish brown color. They typically displayed frontal and mantle white spots, while the females did not. Skin texture also varied with sex; males were papillose while females remained relatively smooth. Mann has also noted these colors in mating giant Pacific octopuses in captivity. Giant Pacific octopuses also used both mounted and distance matings in aquariums and in the wild.

These observations on mating confirm Jennifer's observations on the pygmy octopus, that mating positions may be opportunistic. Some octopus species may mate either at a distance or in the mounted position, depending on their setting, and such flexibility is adaptive. When a male encounters a female in the open while they are foraging, mounted mating may be possible. If a female is in a narrow den, the male may only be able to stretch an arm to her. Since the male is usually the active partner, he probably must do what he can to ensure that contact is maintained (see plate 34).

Observations of matings of giant Pacific octopuses in the wild are incomplete. A complete mating of four hours, including possible courtship or male displays, is hard to witness because of the short time divers can spend underwater before they run out of air, as Cosgrove reported in 2009. Future use of ROVs and submersibles will help us get compete mating reports, but those observations are very expensive, requiring hours of downtime in a submersible.

Mating (not including courtship) is longer in giant Pacific octopuses than in most other octopuses, averaging about 260 minutes. Only the red-spot night octopus has been reported to take longer, 360 minutes. The giant Pacific octopus may need this amount of time because it lives in cold water and all life processes of poikilotherms, which cannot maintain a constant body temperature, are slower in colder water. But James reports that the deep-sea spoon-arm octopus, which lives in water just above freezing, mates in just three minutes, so this may not be the answer. Also, the giant Pacific octopus has the longest octopus spermatophore, each containing about seven billion sperm. Mann states, with tongue firmly in cheek, that this amount of sperm is equal to thirty human ejaculations. Males can easily produce billions of sperm, but it is not easy to produce spermatophores when each contains billions of sperm. The placement of such a large, limited, and valuable resource is highly critical and may take much time.

Jennifer has pointed out that octopus mating is much more than mere copulation, and that documented reports should include other behaviors. Beyond the act itself, there are differences among species in such factors as position, body patterns, courtship behaviors, and male displays. Observations of matings tend to be by chance, but we need more careful studies for comparisons among species. At a minimum, observers should note location, habitat, species, size, relative maturity, position, posture, body patterning and its changes, possible courtship actions or display, contextual variables about possible courtship, mating duration, respiration rate, and the number of the Arch and Pump spermatophore transfers.

After mating, a male octopus goes into a life stage known as senescence. Public aquarium representatives frequently call us about this condition. They say their octopus is not eating and he is losing weight. He's acting strange—not going into his home, and moving around the tank in the open much of the time. As time progresses, his skin develops lesions and the animal loses coordination. People notice and start asking about the welfare of the animal. This phase is a familiar one in the life cycle of the octopus. At the Seattle Aquarium, giant Pacific octopus Clyde weighed 40 lb. (18 kg) when he was collected. He thrived in captivity for five months and grew to 70 lb. (32 kg) but then stopped eating. He crawled all over his tank, even into stinging sea anemones that he normally avoided. He lost 22 lb. (10 kg), 32 percent of his body weight during two and a half months, and then died. Clyde had the normal symptoms of octopus senescence, a precursor to death from old age. These behaviors among octopuses have been observed for millennia. Aristotle said the females became stupid after giving birth, that they could be found being tossed about in the water, and it was easy to dive and catch them by hand.

Senescence occurs at the end of a male octopus's natural life span and often lasts for a month or more. In captivity, however, octopuses may show senescence-like symptoms and even die because of poor water quality (from such problems as lack of dissolved oxygen, low pH, pollutants, or incorrect temperature), stress, and disease.

The physiological processes that drive maturity and senescence in octopuses are now fairly well known. The process is started by hormones from the optic gland, which cause dissolving of proteins in the arms, maturing of the reproductive organs, and eventually inactivation of the posterior salivary and digestive glands and loss of appetite. The optic gland is activated by environmental factors such as light, temperature, and nutrition, which ultimately control timing of reproduction and life span. Wodinsky found in 1977 that removal of the optic gland causes immature females to live to double the normal lifespan but they don't reproduce.

Four conditions or activities are indicators of octopus senescence. First, many researchers and aquarists have noted the loss of appetite in both male and female octopuses at this stage. Second, as the animal loses weight, the eyes of octopuses stay the same size while the body shrinks, causing the eyes to appear to bulge. Third, the increased activity at the start of senescence is undirected activity, not hunting, foraging, or performing other useful activities. Fourth, white skin lesions appear. A young octopus in good condition and with good water quality can get an injury that causes such lesions, but in that case they will heal. But the healing processes of old octopuses cease during senescence, so skin injuries may become secondarily infected.

Both male and female octopuses can go into a physiological decline, but females typically brood their eggs faithfully during this time. Male octopuses go into senescence without a clear transition, but they can survive a surprisingly long time without eating. At the Seattle Aquarium, seven male giant Pacific octopuses stopped eating an average of forty-eight days before dying and lost 17.4 percent of their body weight. In the wild, undirected activity of a senescent male octopus is likely to get him eaten by predators. Senescent male giant Pacific octopuses and red octopuses are found crawling out of the water onto the beach, obviously not a normal behavior and likely to lead to attacks by gulls, crows, foxes, river otters, or other animals. Senescent males have even been found in river mouths, going upstream to their eventual death from the low salinity of the fresh water.

Some observations of octopus couplings have occurred that are puzzling. One bizarre encounter in the depths of the ocean was filmed by the submersible *Alvin* and reported by R. A. Lutz and Janet Voight in 1994. This meeting in the depths would have been interesting just as a film of behavior, since seeing an octopus mating is valuable. But in this observation, the two octopuses were male members of different, undescribed species. The smaller male on top assumed a typical mounted position, parallel to the one underneath. He inserted his hectocotylized arm into the mantle cavity of the lower male and his respiration rate increased.

Such an encounter tells us something about octopus mating systems. If a male detects another octopus, maybe in desperation he'll try to mate with it, with an urge to pass on his genes to the next generation. As yet, we don't know the motivation of the lower male. He was quite capable of turning on the other octopus, and killing and eating him. We don't know the ultimate outcome of this deep encounter, whether the sexually active male realized his error, or whether he simply wasted spermatophores trying to mate with another male.

Octopuses have an intriguing sex life. Hidden beyond what we know may be elaborate courtship rituals, and perhaps there may even be mate guarding, multiple matings, or cannibalism once the act is over. Octopuses die at the end of their reproductive process, females literally wasting away, guarding eggs until they die, and males going through senescence and probably getting eaten by predators. While the octopus's death is not a noble one by our standards, the animal's pattern of reproducing and dying has ensured survival of the species.

11

The Rest of the Group

Scientists are still in the process of learning about the octopus, an animal that is but one example of the class Cephalopoda. The cephalopods are composed of the present-day nautilus, which belongs to an ancient subclass of molluscan cephalopods known as the Nautiloidea, and all other living cephalopods that belong to the subclass Coleoidea. Since the coleoid group has evolved relatively recently, we find it helpful to study the variations among these other animals—squid, cuttlefish, and deep-sea vampire squid—for what they might reveal about the cephalopod pattern. Looking to other cephalopods for similarities and differences in biology and behavior can lead to findings about what it means to be an octopus. So let's examine seven animals whose variations on the cephalopod pattern we know something about.

The first cephalopod of interest is the nautilus, a remnant of the ancestral shelled cephalopods related to the octopuses' molluscan ancestors. The nautiluses flourished at the same time as the ammonoids and belemnoids, early cephalopods that are well known because their fossilized shells are plentiful in rocks up to hundreds of millions of years old. Most of these now-extinct cephalopods died off at the end of the Cretaceous Period, about 65 million years ago, the same time the dinosaurs disappeared. Only one genus of those ancient cephalopods, the six species of *Nautilus*, has survived in the deep ocean, remaining unchanged for at least 100 million years. Nautiluses are considered to be "living fossils," like the crocodile and the horseshoe crab.

Nautiluses live in a coiled shell that can grow to 10 in. (25 cm) across. The shell can be compared with a snail shell, but snails develop a twisted coiled shell while the nautiluses' shell coils on one plane only. And the nautilus shell is made up of many chambers. The animal makes more and more chambers as it grows, and it sits in the biggest, outermost chamber, which is open to the sea. The other chambers fill with fluid, with a siphuncle, a piece of tissue like a living hose, connecting the inner chambers. To change its buoyancy and float up near the surface or return to the deep, the nautilus can vary the amount of fluid in these chambers by pulling it out by osmosis. This technique for buoyancy control is the same one that humans use for operating submarines. In honor of the resemblance, two famous submarines were named *Nautilus,* Captain Nemo's vessel in Jules Verne's *20,000 Leagues under the Sea* and the first atomic-powered submarine.

The nautiluses look like a primitive version of cephalopods in other ways. Their brain is much simpler, though they still have memory like the octopus. Their eyes are more primitive, working like an old-fashioned pinhole camera, with no lens, and open to the seawater. Nautiluses can see images with this system, but their vision is blurrier than the sharp sight of octopuses and squid. Interestingly, they likely use chemical cues in the water to sense when prey is near, in the same way that octopuses get chemical cues from the bottom when searching for crabs and snails. Instead of eight arms, they have about ninety tentacles to grasp their slow-moving prey, usually crustaceans.

Nautiluses use the jet propulsion swimming method like other cephalopods, but it's slower. They take water into the mantle cavity and squeeze it out of the funnel, which is more like a coiled flap. All their activities are done more slowly. In the tropical west Pacific where they live, nautiluses spend much of their lives swimming up or down steep reef faces. They rise at night to feed within 150 ft. (45 m) of the surface, and go down in daytime to around 1800 ft. (550 m) deep, where it is presumably safer for them.

Keeping the flotation balance just right and changing it twice a day must be a metabolic feat. Moving fluid into and out of the inner chambers is part of the process, but nautiluses also actively swim up and down the reef face and swim minimally to maintain their position. If they take out too little fluid, they may sink into the deep, and the excess pressure will implode the shell. If they take out too much fluid, they will float to the surface, bounce around in the waves, and get eaten by predators or washed ashore. Octopuses eat nautiluses; we know because we have found nautiluses' empty shells with drill holes, though it's not clear why octopuses would drill when they could just pull out the animal.

The ability to move down into the deep ocean might account for why nautiluses have survived when so many other early cephalopods are extinct. Their slow metabolism and inactive lifestyle, contrasted with that of the active octopus, fitted them to live in this low-oxygen, colder water. They can survive on one feeding for over two months, so they'd find an intermittent food supply acceptable. Living deep means that they also avoid most marine predators, and they likely find mates and lay eggs in the deep.

Nautiluses are a complete contrast to the octopuses' "live fast and die young" life history. Unlike other cephalopods, they live after laying eggs—they are iteroparous. They lay a few large 1-in. (2.5-cm) eggs at a time. The eggs take about one year to hatch and there's no planktonic stage—the babies just crawl out, already equipped with two or three shell chambers. Like fish but unlike octopuses, they live a long time—fifteen to thirty years. They truly are a transitional life form between the slow snails and the other, fast cephalopods.

The cephalopod that really does live fast and die young is the Caribbean reef squid. Its body is based on the same eight arms and jet propulsion with the mantle, though it has an additional pair of elastic, elongated tentacles for prey capture. True squid live in groups of hundreds and swim constantly and fast in the surface waters of the open ocean, but the Caribbean reef squid is unusual in that it lives in smaller groups and nearer shore. These squid are big for a cephalopod, 1 ft. (1/3 m) at adulthood. They live about a year or a year and a half, again depending on how cold the water is where they live. The eggs are relatively big (about 1/3 in. long, or 2/3 cm), and the tiny, newly hatched squid drift in the open water, eating small crustaceans and fish larvae. As they get older, they take larger prey, more fishes than other groups, swimming close to a school of small fish and shooting the paired tentacles out to catch them.

Because these squid live in a nearshore habitat, we can study their behavior more easily than if they were far out to sea like their relatives. Science has to bend to practicality, which is why open-ocean and deep-sea animals are so poorly known and why much of our knowledge of cephalopods such as the vampire squid is speculative. Martin Moynihan and Arcadio Rodaniche (1982) spent some of several summers watching the Caribbean reef squid at the San Blas Islands, tracing their life history and suggesting that they make a visual language on their skin. Female squid don't usually tend and defend their eggs, and the Caribbean reef squid lay theirs in 100 strings of about three eggs each, hiding them under rocks or dead coral where scavengers won't find them. Young squid probably move into similar habitat as the adults, though scientists have also found them in sea grass beds. They swim together in the daytime and disperse at night to catch small crustaceans and fish. Juvenile squid spend the daytime just hanging around together doing nothing much. But when they grow up, choice, competition, fighting, and courtship become the focus of their days.

Jennifer has studied these squid for years on the island of Bonaire, a Dutch island in the southern Caribbean and part of the Netherlands Antilles, whose great attraction is that all the shoreline of the island is a protected marine park. The island is informally known as a divers' paradise for its easy access to dive sites and abundant underwater life. There, groups of squid swim together up to 120 ft. (40 m) from shore in crystal-clear water less than 15 ft. (5 m) deep. Jennifer has been exploring chromatophore form and function in squid, and whether squid make a visual language on their skin, which Moyhihan and Rodaniche have suggested. Many octopuses and squid have such a phenomenal range of colors and control of changes in skin texture and pattern that if any cephalopod group has a visual language, it would be squid.

Adult Caribbean reef squid court and fight using visual patterns on the skin. There's the female saddle and the male stripe, exchanged in early courtship to indicate interest, and the highly visible on-and-off male Flicker, showing a drive to mate. But the aggressive zebra display (see plate 35) is possibly the most interesting and definitely the most variable of the displays. It's a pattern of dark diagonal slashes, often on the mantle and sometimes on the arms, made mostly by males.

We have seen squid displays used only for camouflage or courtship. So far, we think squid don't have a language, although they have a fascinating and variable skin display system. Their skin system could make patterns complex enough for a language. Maybe they haven't made a language because of their relatively simple social system. Squid only have a burst of testing, consortship, and competition at the end of their lives. They haven't evolved a language because they don't have enough to say with it.

Like the large reef squid in structure, with its elongated mantle, eight arms, and two tentacles, the tiny pygmy squid (*Idiosepius pygmaeus*) is quite different in life history. The primary difference is size. Adult pygmy squid grow to about 1 in. (2.5 cm) of mantle length, no more. They are about the same size and shape as little grass shrimp, they live in the same sea grass habitat, and they move with the same short, jerky motions that shrimp accomplish with tail flips. They are so different from the other squid that they have the whole order Idiosepiida for just the one genus. In fact, because of their small size and less linear shape, the pygmy squid were misplaced for a long time in the Sepiolidae, with the cuttlefish and the stubby squid.

We don't know much about pygmy squid, probably because the group lives in the Indo-East Pacific. Pygmy squid live quite near shore, probably in very large numbers, mostly in shallow beds of sea grass but sometimes near or on floating vegetation. They have a short lifespan of around 100 days. Their young are planktonic, but as adults they move around little. The most interesting thing about their ecology is that the larger adult females (see plate 36) lay egg after egg after egg in rows along the blades of grass. For this large output, they are "minimaximalists," miniature in size but maximal in egg production. The females spend a lot of time doing this; they will lay a row of eggs, go back to feeding, and then lay another row for up to two weeks, 15 percent of their lifespan. Meanwhile, males display, fight, and court around females, trying to mate. Neither sex makes exclusive pairs with another individual, nor do males guard females, like reef squid do. Females can attract males with a color and posture display, and males can do the same for females. We have seen two or even three males grab onto a female as she laid her eggs, passing spermatophores to stick below her arms as she went on unconcernedly putting down more and more eggs.

We know very little about pygmy squid behavior. They have a fascinating habit of attaching onto grass blades with a sticky glue gland on the upper surface of the mantle, and they can detatch and swim off in an instant if threatened. But we don't know how they get attracted to the grass beds when they are young and drifting in the plankton. About their foraging habits, we know that they eat small crustaceans. For people who want to study these fascinating squid, there's a full bibliography in von Boletzky's 2003 report.

It seems logical that, like pygmy octopuses, pygmy squid would be easy to keep in an aquarium, but this has proved wrong. They are tiny when the eggs hatch, but so are giant Pacific octopuses. They seem to tolerate variations in temperature and salinity that might occur in an aquarium. We don't know what immature pygmy squid would eat; since they live in the plankton, maybe an enterprising squid owner would have to go out with a boat and do plankton tows, very difficult if you don't live near the sea and have access to boats. There are lots of good questions to ask about this cephalopod minimalist, and raising them in the lab would help find answers.

The giant squid is at the other end of the squid size range and is the biggest cephalopod. While it seems like an extreme match, giant squid are relatives of the reef squid and pygmy squid. They are perhaps the best publicized and least known of all the cephalopods. Many people have read about them in adventures such as *20,000 Leagues Under the Sea*. Books about this squid have been written, but even the 1998 authoritative volume by Richard Ellis, *The Search for the Giant Squid*, has only a slim chapter on what is actually known about them.

We know that giant squid get big, and we know sperm whales eat them and are their main predators, but we know little else of their lifestyle. There are no authenticated cases of giant squid ever harming a human. No boats have been pulled under by them. No humans have been eaten by them or even been pulled underwater by their long tentacles. Few have ever seen a live adult giant squid, despite sailors' tales and several well-funded expeditions to find one using manned submersibles, ROVs, and even cameras mounted on sperm whales. Tsunemi Kubodera and Kyoichi Mori (2005) were able to lower a camera into the depths and get video of a giant squid far, far below.

What we know about the giant squid comes from examining their carcasses washed up on beaches, trawled in nets, found floating on the surface, or dissected out of sperm whales' stomachs. The first giant squid was described in 1545 off Scandinavia. It wasn't named until Japetus Steenstrup described its genus in 1857 as *Architeuthis*, which means chief squid, and named two species. Since then, fifteen other species have been described. But the systematics of the group are hopelessly confused, like for many other cephalopods. Some researchers believe there is only the one species worldwide, *A. dux*.

How big does the giant squid get? Not as big as legends suggest, since the extended tentacles are added to the length measurement when describing its size. The largest one, measured at 55 ft. (18 m) long, was found off New Zealand in the 1880s. Its body was only 8 ft. (2½ m) long. During the twentieth century, the largest squid found was 47 ft. (15½ m), at Grand Bahamas Bank. A number have been found recently in the New Zealand area; all were about 25 ft. (8 m) and weighed about 500 lb. (227 kg). Most had nothing in their stomachs so they may have been of reproductive age or senescent.

Most scientists today believe that despite its size, the giant squid is not a threatening animal. But this belief is speculation; we have few hard facts about the behavior of the animal. It has ammonia in its tissues for buoyancy, its mantle is not all muscle, and the ammonia means that it tastes terrible. It has weak muscles in its mantle and small fins so it probably is not a fast swimmer. It may hang in the water with its arms and tentacles dangling down, waiting for a fish to swim into them, which is unlike the octopus's active foraging.

The giant squid has chitinous toothed rings around the outer edge of each sucker on the arms and on the tentacle tips. Other squid also have hooks on the arms and/or suckers but the giant squid does not, which also suggests that it is not the fierce predator that fiction would have us believe. The sucker rings help the animal to grasp prey, and they also leave circular scars on the skin of the sperm whales that eat them. Old-time whalers spoke of squid scars the size of a dinner plate, leading them to speculate there might be huge monster squid lurking in the depths that haven't yet been seen. But no sucker scars larger than 2 in. (5 cm) across have been documented on any whales. For a comparison, a 95-lb. (43 kg) giant Pacific octopus at the Seattle Aquarium had suckers 3 in. (8 cm) across.

Although the war between giant squid and their sperm whale predators has been depicted as a battle, this appears not to be the case. There is no evidence beyond the sucker scars that sperm whales ever have much trouble eating a giant squid. Mariners have seen sperm whales bringing giant squid to the surface to eat them. Such reports typically recount that the whales were methodically munching their way through the squid with their muscular toothed lower jaw and swallowing them with ease.

It's quite a jump from the fast, sleek, open-ocean true squid to the bottom-living and much slower sepiolid squid, which resemble cuttlefish. On the surface, they look much like octopuses, but there are big differences. Many sepiolids are nocturnal and so small, 2 in. (5 cm), they are easy to overlook, even though they are very common.

The stubby squid is not a true squid but a sepiolid, but it's otherwise appropriately named since it is very short. The sepiolid squid are one of a group that lives mostly on or near the bottom. They are mostly small, less than 4 in. (10 cm) in mantle length, and live less than six months. They look like a cross between an octopus and a cuttlefish, with large round eyes, a fin on each side of the body, and two long prey-catching tentacles like squid. Stubby squid range around the rim of the North Pacific in water down to 3000 ft. (1000 m) deep. They feed mostly on shrimp, which they catch with their two long tentacles. They have a small beak that they use to chew shrimp like we eat corn on the cob, eating them from the center toward each end, consuming most of the tail meat and the internal organs from the body of the shrimp and leaving the shell. They are alert at night, when divers see them sitting on the sand or mud bottom watching for shrimp, predators, or mates.

To avoid getting seen during the day by visual predators, the stubby squid buries itself in the substrate, using a fixed action pattern—a set of the same behaviors used in the same sequence each time—for burying. It sits on the sand and blows underneath, first forward with its water jet then backward, and continues blowing forward then backward alternately until it creates a depression in the sand where it sits. Then it reaches out on both sides at a 45-degree angle with its second pair of arms, scooping up a bit of sand, pulling it close, and then throwing it on its body and head. It repeats this throwing action until it is covered with sand.

Once covered, the squid consolidates the sand next to its body with thick mucus from the skin and pokes its eyes up through the sand. It aims its funnel straight upward and blows one or two strong water jets out to clear a breathing hole. Then, keeping the funnel pointed upward, it breathes more shallowly than the octopus, taking water in through the gill slits and out the funnel. It sits in this mucus-bound cocoon through the day, protected from predators by being almost totally out of sight.

One of the largest of the cuttlefish group, and well known because it has been hunted by humans for food for centuries, is the common cuttlefish of European waters. Full size at 1 ft. (1/3 m) in length after a lifespan of one and a half years and living near shore over sandy mud bottoms, the cuttlefish is a valued food item. Like the octopus, it lives near the sea bottom but one clear difference between the octopus and cuttlefish is buoyancy. Cuttlefish have a wide calcareous cuttlebone just inside the dorsal surface of the mantle. Its many chambers all have air inside them, which helps the fairly heavy-bodied animal to keep buoyant. The cuttlebone is given to caged birds for them to nibble at and take in calcium. The cuttlefish is a bottom dweller like the octopus, although it does not need to be, and even sometimes buries itself in the sand. In the daytime, like stubby squid, it is almost invisible under a sand layer.

Even though they can float all the time, cuttlefish are like octopuses in that they spend much of their time hiding in the landscape. But cuttlefish also have a truly great repertoire of colors and patterns on the skin. Cuttlefish skin patterns have the same eyespot pattern that squid have,

Angling for Dinner

Squid may sleep while buried, but they are usually aware of their surroundings even if their eyes appear to be buried. I tested their visual alertness when they were buried by moving my hand toward the tank without actually touching it, and almost every time there was a reaction. At first the squid would blow a jet of water straight upward with sand particles, so it looked like a little sand geyser erupting from the bottom. Repeated threat gestures caused the squid to next blow out a diffuse cloud of ink and then a thicker ink blob, definitely a giveaway that a creature was there. Finally, the squid would emerge from the sand with a sand coat on top of it, blow out an ink blob, turn pale, and jet away. The thick ink blob is about the same size as the squid, and may confuse predators.

While I was studying the burying behavior of the stubby squid, I saw them use another curious behavior that looked suspiciously like "angling." A squid would partially bury itself, so that its eyes were well above the sand. It would turn its body and head a dark rich red-brown, then poke the arm tip of one of its first pair of arms up out of the sand. The squid then wriggled the arm tip erratically in front of one of its eyes. This action may attract shrimp or small fish, which the squid can then catch.

Specialized angling behaviors are well known from fish such as the anglerfish, which dangles a lump of flesh at the end of a muscular rod. It waves the lump like the bait on the end of a fishing line at hungry smaller fish. When they come close, it quickly grabs them in its large mouth. Other cephalopods also use angling. Cuttlefish and Caribbean reef squid use an arm-waving behavior, possibly toward potential prey, in combination with differences in color and body posture. This behavior may occur in the four major orders of the coleoid cephalopods and is another way they could change their appearance to survive, but it hasn't been systematically studied. Luring with the tip of an extended arm could take place in octopuses too, an example of the tremendous flexibility of the foraging strategies in the group.

—Roland C. Anderson

they have dapples and dots, and they make zebra stripes as an antagonistic signal. It's even rumored that if you put a small cuttlefish onto a black-and-white checkerboard pattern, it will make the same checkerboard on its skin. Keri Langridge et al. (2007) are beginning to separate out the various postures and patterns that cuttlefish make to a potential predator (also see Hanlon and Messenger 1988). Having this repertoire of skin patterns and colors may sound like a paradox, since the cuttlefish is usually buried under sand all day and is only active at night when there isn't much light for anything to see these patterns. Maybe, like the octopus, it comes out sometimes in the day. Or maybe there was huge selection pressure to get the appearances right even though they weren't used very often. Or maybe cuttlefish can't always find sand.

When they are out of the sand, cuttlefish are generalist predators like octopuses, but the prey species they take and the way they catch them are different. Like true squid, cuttlefish have a pair of elastic, extensible tentacles. These can be shot out, just like chameleons flip out their tongue, to grab small prey, often assisted by suckers on the tentacle tips that make sure the prey stays caught. And cuttlefish are flexible in capture technique: if a prey animal such as a crab is slow and too big to catch with the tentacles, they will just grab it with the arms. They may also blow sand off buried crabs and use their tentacle tips as a lure. And cuttlefish have a distinctive feature: they nearly always use vision to help them find prey, and there's no exploring in the rocks and crevices like the octopus. Not much of the cuttlefish brain is allocated to processing touch information.

Compared with the common octopus, there has been almost no field observation of behavior of the common cuttlefish. We don't know why, since the cuttlefish is a very common species in European waters and a valuable food resource. Cuttlefish live near shore, and though they often are found in muddy water, they are certainly big enough at around 1 ft. (1/3 m) mantle length to watch easily. They have been raised in captivity, and a lot of work has been done in the lab on their early behavior, as well as on their reproductive interactions, much by Boal (2006).

Cuttlefish probably have much more interesting sex lives at the end of their one-year life spans than octopuses, based on lab work and field observations of the giant cuttlefish in Australia. Even when they are mature, most octopuses don't seem to encounter each other a lot, but cuttlefish often migrate to spawning areas where males compete for access to females. Males give each other zebra stripe skin displays and even engage in physical battles. In the restricted environment of the lab, they corner females, mate with them head-to-head, and even try to blow the sperm of previous males out of the females' oviducal gland with a jet of water. Females sit back and decide which male to accept. They may use chemical cues to guide their choice. Boal set up choice tests in 1997, and found that females chose those males that had successfully mated with other females, not just the biggest males. Unlike in the octopus, there is no care of eggs in cuttlefish. The eggs are attached to rocks or algae by females, and then both sexes die.

On a small rocky reef in South Australia in 2002, Katrina Hall and Roger Hanlon observed a huge aggregation of giant cuttlefish (*Sepia apama*) ready to reproduce. Such mating aggregations can be spectacular, like a cuttlefish singles bar. Females stay for a short time, and so they may have just arrived, because there were four males around for each female. Big males guarded or tried to guard a female, though without long-term success, because pairings only lasted ten to fifteen minutes. Bigger ones always won contests. Smaller males waited around until a guarding male was distracted and then tried a quick courtship of females. They would take on the skin patterns typical for a female and hang around her without the guarding male noticing. They probably didn't fool her, though, as over 70 percent of all mating attempts by males were rejected and females had a specific skin display to indicate they weren't interested. With over 41,000 giant cuttlefish in the area and a density of about one per square yard, it must have been a sight to see. The common cuttlefish probably does something like this though in smaller groups.

Hardly anyone knows much about the vampire squid, but two things stand out: it's very different from other coleoid cephalopods, and it's well adapted to the harsh environment of the ocean depths. Taxonomists worry about where to put *Vampyroteuthis*, referred to as the vampire squid (even though it's not a squid) because that's what the name means. Some taxonomists place it in a separate order, the Vampyromorpha, within the Octopods, and some put it in a separate group completely.

The systematists put the vampire squid halfway between the squid and the octopuses on the family tree. Two things stand out about the vampire squid: it's very different from its octopod and squid relatives, and it's well adapted to the depths of the oceans in which it lives. Vampire squid differ from other coleoids in several ways. They have peglike cirri instead of suckers on their arms like some of the deep-sea octopuses, and they also have fins like the squid do, though the fins are elongate and attached to the posterior instead of attached laterally like in the squid. Vampire squid have no chromatophores except near the photophores, or light organs, so they are poor at color changes. They have lost the ink sac and the specialized hectocotylized arm. Most unusually, they have a pair of long filaments that are like tentacles but much more fragile, usually kept tucked in pits between the first two arm pairs. They trail these long filaments through the water as they move, trolling for prey in the same way we might troll for salmon. When a prey species hits a filament, the squid will circle back to pick it up with the arms. These structures aren't a standard set for coleoid cephalopods, but the lack of chromatophores and the presence of filaments and cirri make sense when we realize that vampire squid drift in the dark, deep sea, between 1800 and 3000 ft. (600 and 900 m) down. Because vampire squid live at such depths, many of our ideas about how they live are guesses. Until ROVs started taking videos deep in the ocean, scientists were basically working on preserved bodies of animals that had been trawled up from the depths.

The deep ocean seems like a harsh place to live. There is unimaginably high pressure: if you dangle an empty coke can to the bottom of the ocean and then bring it up, you'll get a small crumpled ball of metal. Living in that pressure may be less challenging for soft animals than for armored ones, since pressure changes could crumple skeletons. And high pressure may interfere with metabolism in unknown ways. Then there's the water quality: plants in the deep do not use sunlight to help make food and give off oxygen, so the water contains a low percentage of oxygen. It's also cold, a constant few degrees above freezing. Fewer species can survive in this inhospitable environment, and finding prey may be very difficult. And there is little to no light available.

Deep-sea and cave animals adapt to the lack of light in two different ways. One way is to evolve into not using light as a sensory cue at all. Like the cave crayfish in Florida, some species have reduced eyes or none, having gone completely blind. The deep-sea blind cirrate octopus (*Cirrothauma murrayi*) has done just this: the eyes are tiny, pigmented cups and there's no lens or any other focusing device. They can detect light but that's about all—no visual shapes, no object recognition. To compensate for the lack of light for vision, they use other information sources. Cephalopods already have the ability to use and learn from touch information, and no doubt the cirri on the arms of vampire squid are very sensitive to touch. This ability is shared among the true squid, which have a row of pits along the edge of the body, each containing a mechanically sensitive receptor. But the vampire squid's filaments extend the range of touch many body lengths, and anything triggering one of these feelers can be attacked by the squid. Probably the vampire squid have excellent chemical sensitivity, but we don't know that yet, because we've never been able to keep an animal of this species alive in the lab.

A second way to deal with the lack of light is to make one's own light by creating bioluminescence. This feature is common in many shallow-water, night-active animals, like the sea pen and the lanternfish, which have to cope with darkness. Vampire squid have also adapted to the dark: they have wide-pupil blue eyes that pick up the maximum amount of light. They have a variety of devices to make their own light and the techniques to manipulate them. First, although they do not have an ink sac, vampire squid have a bioluminescent mucus that they can jet out, presumably at the approach of a potential predator, likely distracting it in the same way as a black ink jet for a shallow-water octopus or squid. Second, they have a pair of light organs at the base of the fins with a moveable flap that could be used as a shutter. These could act as a searchlight, turning a beam of light onto a potential prey species that tactile sensing from the filaments has picked up. And third, they have a huge number of tiny photophores all over the body and the arms. These could work two ways: they might give a general dim lighting as visual counter-shading. With even a little light from above, a dark animal would stand out in silhouette from below. With low-level light giving just enough illumination, it could blend in. And the second function of these lights has been seen by ROV viewers: a disturbed vampire squid threw its arms back over its body and flashed the lights on the arms, which should startle any creature.

Vampire squid deal with the cold, dark, oxygen-poor water around them by slowing down their metabolism. They glide through the water, flapping their peglike fins and maybe contracting their arm webs as jellyfish do their bell, though they can make bursts of speed. Slow movement would mean they could live with less oxygen—back to the basic molluscan model—which allows them to live in places where fast but oxygen-demanding predators can't live. Cold would also slow down vampire squid. We don't know whether cold would also lengthen their lifespan, but it's a good bet, since James found that the deep-sea spoon-arm octopus lives at least six years. By the way, for those imagining that vampire squid are monsters of the deep, they are tiny—only up to 5 in. (13 cm) long.

A disadvantage of the sparsely inhabited deep is that when it comes time to reproduce, vampire squid may have trouble finding a mate. Perhaps vampire squid use chemical cues, or they may flash coded messages to each other with their photophores. One possible clue comes from the size of the vertical lobe in their brain, which is the learning and storage center for visual information in octopuses. The vertical lobe in vampire squid is a huge percentage of the brain size, a whopping 28 percent, compared to the common octopus's 13 percent. What is that large lobe needed for? This vertical lobe is the learning area in octopuses. The bigger a brain area, the more important the function that's controlled by that area. It's not surprising, for example, that octopuses have a large brachial lobe devoted to control of all its arms, and the true squid have a smaller one.

The deep sea is such a difficult environment to live in that extreme adaptations have to be made to the basic coleoid plan. That is a lot of evolutionary pressure, though. The many octopuses without cirri, true squid and sepiolids, thrive in shallow water and among fierce competition. Of the huge number of species in the Octopodidae, only a few species—the vampire squid, the finned deep-sea octopuses, and the glass octopus—have managed what must have been a difficult transition to this demanding habitat.

This glimpse into cephalopod variations leaves us with a lot of questions. Why is the pygmy squid so small and giant squid so big? Which are better at camouflage, cuttlefish or octopuses? Which cephalopod is likely to become social, and will that take its intelligence in a whole new direction? Why does the vampire squid have such a huge vertical lobe in its brain, and does brain lobe size indicate the amount of information processing that a particular brain area handles? Why are most cephalopods semelparous—having one reproduction at the end of their lifespan—and others like pygmy squid and nautilus are quite different, continuously laying eggs once mature. How does the octopus control all those arms and coordinate all those suckers, and will learning about squid arm actions help us understand octopus arm movement? It will take a new generation of octopus and squid researchers to investigate these questions.

Still, the three of us have helped lay the foundation for the recent study of octopuses. In our work, we have combined three aspects of science—basic research, practical applications, and communicating discoveries to the public. All dimensions are important, and interweaving them is the way to do science. Jennifer's observational field research from the 1980s is still being cited. Her comparative work, often with Roland, on personality, play, and tool use challenges conventional thinking about invertebrates. Roland has decades of experience with captive octopuses, and with James is an author on one of the world's first papers to propose applying the concept of enrichment to all animals: giant Pacific octopuses were used as model organisms in that publication. In 1995, James embraced the World-Wide Web as a forum for sharing research on the behavior, life history, husbandry, and physiology of cephalopods with an appreciative, varied public.

Maybe our most important accomplishment is two-fold. We have shown what many suggested for years, that the octopus is a very smart animal. Then, we've moved forward in examining the ethics of treatment of this invertebrate and in developing practical ways to assure that octopuses are well cared for. And we are still having a great time doing it.

Postscript
Keeping a Captive Octopus

Octopuses can be kept as pets. They are occasionally sold at aquarium shops, for this reason. With a few exceptions, related to laws in particular states and a few rare species (which don't live off North America), collecting a local octopus from the ocean to keep in an aquarium is sustainable. The effort required to catch a single local octopus helps to prohibit unsustainable exploitation. With proper knowledge and equipment, advanced aquarists can successfully keep healthy octopuses in captivity. But these remarkable animals have some special needs. We only recommend keeping octopuses to those who are already familiar with maintaining marine (saltwater) aquariums and have knowledge of the octopus's ecology, life history, evolution, and behavior.

Octopuses watch us as much as we watch them. It is sometimes hard for us to remember that these invertebrates are mollusks, with relatives such as slugs, snails, clams, and oysters. After all, clams and oysters don't have a brain or even a head; they simply have a hard shell to hide within. Cephalopods—octopuses, squid, cuttlefish—are often compared with fish, the dominant group of large vertebrates in the ocean. Like fish, they have well-developed senses and brains, are active, and are found in all oceans of the world. A wonder of the natural world is how similar the eyes of an octopus and a fish are, even though they are two unrelated groups of animals. But in other ways, cephalopods are very different from fish. Most fish, with some exceptions like salmon, live many years and reproduce multiple times. If the reproductive output in one year is low, the population of mature fish will be around the following year to reproduce again. Because most octopus species only reproduce once, they literally put all their eggs in one basket. The combined effect of this with their short life span of only one year or so makes octopus populations incredibly sensitive to small environmental changes. In tough years, populations can crash.

The Challenge

The short natural lifespan of most octopuses doesn't change, even under the best care in captivity. Animals collected in the wild have already lived a large part of their short lifespan. It is difficult to know how old an octopus collected in the wild is; even size is a poor indicator. In the laboratory, octopuses the same age, with the same genes (siblings), raised at the same temperature, and fed the same food can easily differ in weight by a factor of ten or more. In the wild, where genetic composition, access to food, and temperatures differ, one could expect even greater size variability at the same age. Anyone considering keeping an octopus in captivity needs to understand that with care, octopuses are not likely to live much more than a

Best Choices for Captivity

These three octopus species generally do well in captivity. The first is often found at public aquariums, and the other two can be kept by aquarists. All three are relatively well studied and are not likely to be overfished.

Giant Pacific Octopus

Size: Huge, up to 100 lb. (45 kg) in captivity. This octopus is too big for all but the largest aquariums. As a mid-sized adult, it should be kept in a tank no smaller than 500 gal. (2000 l).

Life history: Lives approximately 3 to 4 years at a temperature of 46 to 54°F (8 to 12°C). Lays tens of thousands of eggs. The tiny offspring, each 0.0009 oz. (0.025 g) spend the first part of their life in the plankton.

Habitat and range: Can be found off the coasts of the northwest US, western Canada, Alaska, Russia's Kamchatka Peninsula, and Japan. Their distribution is restricted to relatively cold water.

Water temperature: 46 to 54°F (8 to 12°C). Maintaining these low temperatures typically requires an open system with a cold-water source or the use of a high-capacity chiller.

Behavior in captivity: Can be cannibalistic. This species has a reputation for being talented in escaping captivity.

Caribbean Reef Octopus

Size: Mid-sized octopus that typically weighs up to 35 oz. (1 kg) as adults.

Life history: Lives approximately one year. Should be kept in systems that are 30 gal. (114 l) or larger. Lays approximately 500 large eggs, which hatch into benthic nocturnal offspring.

Habitat and range: The Caribbean Sea. This species is the most abundant in the shallow waters around the Florida Keys, although many people never see them since the octopus is strongly nocturnal. They can be found on rocky intertidal areas and in sea grass beds, but the best place to look for them is rock walls and jetties. These octopuses are most abundant in shallow water, 3 ft. (1 m) or so deep.

Water temperature: 77 to 82°F (25 to 28°C).

Behavior in captivity: This species typically learns where and when the food comes from in captivity and adapts its behavior accordingly. After a few months, they will often plaster themselves to the front of the aquarium at feeding times. The species is one of the more cannibalistic. They are also quite adept at escaping. These octopuses have some beautiful nocturnal color patterns, using iridescent blue-green colors from their irrodophores sometimes combined with a reddish marbling.

Caribbean Pygmy Octopus

Size: These small octopuses weigh less than 3.5 oz. (100 g). A 10-gal. (38-l) tank is sufficient space for one animal. But since smaller marine systems are less stable than larger ones, we suggest connecting a small octopus tank to a larger system.

Life history: Pygmy octopuses only live six months to one year, depending on water temperature. There are two sister species of pygmy octopus: One species lays approximately fifty eggs, which hatch into direct-developing offspring, which on average weigh 0.0014 oz. (0.04 g); the other species lays more but smaller eggs and has a planktonic stage.

Habitat and range: These octopuses are often collected in the Florida panhandle region in and around oyster reefs. They are likely all over the Caribbean region, and are found in shallow water 3 ft. (1 m) to 33 ft. (10 m) deep.

Water temperature: 68°F (20°C).

Behavior in captivity: These small octopuses are strongly nocturnal and often remain that way in captivity, so they may hide all day. They are not very cannibalistic, are hardy, and have only a mid-level likeliness to escape.

—James B. Wood

year, often less, before dying from natural causes. Giant Pacific octopuses live for about three years, but you'll need a swimming pool–sized tank to keep them in. Those who choose to keep these animals must do everything they can to ensure that proper care is given during this brief period, so that the octopus survives to its natural old age.

Octopuses are different from most animals kept in captivity. Although they are mollusks, in many ways they are not like other mollusks. They live and compete with fish. They have evolved into a different niche—that of a fast-growing, short-lived marine predator. Amazing animals, octopuses do not easily fit into what we typically think of as a mollusk, so they are sometimes referred to as aliens of the sea. Understanding their life cycle and behavior in the wild can help one to make informed decisions about keeping one in captivity. Many people believe that keeping octopuses is extremely difficult, but with proper knowledge and marine aquarium experience, it is doable and fun.

Finding an Octopus

Obtaining an octopus is often a big challenge, especially if you live far from the ocean. There are several methods for getting an octopus for an aquarium. For most hobbyists interested in keeping a pet octopus, the easiest way to obtain an animal is to buy one at a local aquarium store. Even if they don't regularly stock cephalopods, some saltwater shops will special order an octopus. By going through a pet store, you can reduce the cost. If the octopus is already in the store, you can make sure the octopus is feeding.

To establish the health of an octopus, the easiest way is to observe whether it has a good appetite. Because pet stores buy through distributors and in bulk, a cephalopod at a local shop may cost about half what it would if you obtained it yourself through a mail-order company; the required overnight shipping is expensive and saltwater is heavy. And if the shipment has to cross a national border, it can become mired in red-tape permits and sometimes blocked by bureaucrats. When you purchase an animal through a pet store, you can examine it and ask the shop owners questions before you take it home.

When asking questions, keep in mind that pet shop owners often lack specific knowledge about octopuses and their care requirements. The online shops and mail-order houses usually know very little, too. For example, many suppliers label all species of octopuses as the common octopus regardless of where they came from or the actual species. The animal you receive could be a baby with the potential to grow very large or it could be a full-grown adult of a small species that may have only weeks left in its lifespan. But most octopuses in the aquarium trade are in the small to medium size range as adults—rarely larger than a grapefruit.

Transporting an Octopus

Scientists and hobbyists alike are frustrated when transporting octopuses long distances is left to others. In addition to costing more than many marine fish and invertebrates, octopuses don't tolerate shipping stresses well, and there's always the chance of the animal inking during shipment, and the ink can kill an octopus. We've been told that almost any type of cargo has higher priority than live fish or invertebrates. Some octopus species are only available in certain seasons. And prices do not include shipping costs, which can be over fifty dollars for the smallest specimens and thousands of dollars for the largest.

If you live near the ocean and are comfortable in it, collecting an octopus from the wild is usually the best method. With some practice, you can select an octopus based on condition, size, and species. Be sure to check local fishing and other regulations before collecting any marine life, or you might receive a hefty fine. In Bermuda and the Caribbean, we often catch octopuses for research projects, mostly common octopuses (see plate 37), which can be found in shallow water while you are snorkeling. Snorkeling allows more maneuverability and more bottom time than diving. We use a diver's tropical fish-collecting net, which has an aluminum frame with heavy clear plastic on the sides and a screen bottom. These nets are designed to minimize stress on marine life: the plastic sides don't abrade an octopus's delicate skin like a coarse mesh net would. And these nets allow a collector to briefly transport the octopus out of the ocean while keeping it submerged in water in the plastic part of the net.

The trick to using these nets is to resist the urge to chase the octopus with them. While chasing after butterflies with butterfly nets works on land, divers and snorkelers are too slow and bulky to catch mobile marine life in their natural environment. Instead, place the net strategically on the bottom with the opening facing toward the animal you wish to catch, one side resting on the bottom and another against a large rock. Hold the net as still as you can with one hand. Then, with your other hand, coax the animal toward the net. Sometimes you have to use a tickle stick or your hand to get the octopus to swim or crawl into the net. Any long, thin stick up to 3 ft. (1 m) can be gently used as a tickle stick. In Florida, yellow tickle sticks are commonly sold in dive stores during lobster season. You can carry a net in one hand and the tickle stick in the other. Use the stick to reach

Warning: The Deadly Blue-Ringed Octopus

Blue-ringed octopuses, the four species that are members of the genus *Hapalochlaena*, display stunning coloration. Like other spectacular forms of marine and terrestrial life, they have vivid color patterns as a warning signal. These small octopuses pose a serious threat to humans. They pack a potent venomous bite that makes them among the most dangerous creatures on Earth. Their venom, the neurotoxin tetrodotoxin (TTX) described by D. D. Scheumack et al. in 1978, is among the few known cephalopod venoms that can affect humans. A variety of marine and terrestrial animals produce TTX. Terrestrial examples include poisonous arrow frogs, newts, and salamanders. Some marine invertebrates, such as xanthid crabs and a sea star, can produce TTX, but the classic example, and what the compound is named after, is the tetraodon puffer fish. The puffers are what the Japanese delicacy fugu is made from. If the fish is prepared correctly, extremely small amounts of TTX only cause a tingling or numbing sensation. But if it is prepared incorrectly, the substance kills by blocking sodium channels on the surface of nerve membranes. A single milligram, 1/2500 of the weight of a penny, will kill an adult human. Relative to other venoms and poisons, by weight TTX is 10 to 100 times as lethal as black widow spider venom and more than 10,000 times deadlier than cyanide. Even in the minuscule doses delivered by a blue-ringed octopus's nearly unnoticeable bite, TTX can shut down the nervous system of a large person in just minutes; the risk of death is very high. This risk associated with keeping blue-ringed octopuses, plus their short life span, make them a poor choice for a pet.

—James B. Wood

into places such as an octopus den, where you might not want to place your fingers, to help persuade its resident to leave. A tickle stick can also be used to extend your reach when trying to get an octopus to turn around when it is going away from the net.

Some collectors use chemicals to assist in catching octopuses (Roland is very wary of this practice because of the risk of damage to the octopus). Chemicals are used to persuade an octopus securely hunkered down in its den to come out. You can use a large syringe without a needle or a turkey baster filled with a chemical that is annoying to octopuses but not dangerous. Bleach, ammonia, or alcohol is often used for this purpose. The chemical is squirted deep into the den, and the annoyed octopus usually quickly vacates.

Once you have collected an octopus, the next step is to transport it to its new home. With oxygen and a little luck, octopuses can survive up to twenty-four hours in transit if they don't ink, and longer with water changes. Anything over this is pushing it unless there are water changes and near constant attention. Cephalopods simply do not ship well.

Shipping Octopuses

As a graduate student in Canada, I ordered cephalopods from collectors in Florida and California on several occasions. Survival rates were low due to the long shipping time. The worst shipment entered Canada in Ontario two days after being sent from an aquarium wholesaler. The airmail package was then placed on a truck for the trip to Halifax. It arrived eight days later. Needless to say, the package was not a joy to open.

While at the University of Florida, I would go to the Florida Keys to collect octopuses. These spent up to three days in buckets and later in transit back to the University. But I paid close attention to them: I used standard 110-volt bubblers while in the Keys and battery-powered bubblers during transit, did frequent water changes, monitored the water chemistry, and in general made it a priority to provide the best care possible. I have never lost an animal I collected myself.

—James B. Wood

Personally taking a round-trip airplane ride may sound like overkill for obtaining an octopus in good condition. But for research scientists wishing to obtain a number of animals from a thousand miles or more away and ship them across international borders, it is often the most economic and viable solution. By being there, you can ensure that the octopuses have the best care during shipping and that they proceed through bureaucratic snags with the most haste possible. John Forsythe, formerly of the National Research Center for Cephalopods (NRCC), flew to Japan, Australia, and other locations to collect, pack, and escort cephalopods or their eggs back to the NRCC. In his expert care during every step of the trip, the animals had a much greater chance of survival.

In the cold Pacific waters off of Washington state and under the direction of Roland, giant Pacific octopuses and red octopuses are collected by Seattle Aquarium scuba divers. Timing and choice of dive site are important, since octopuses, particularly giant Pacific octopuses, are not seen during every dive. Collecting red octopuses by diving is usually easy, because this small species is found frequently during the day inside discarded beer bottles. At night they are even easier to find since they are crawling about. During the dive, the animals can be placed in plastic zip bags and then transferred to a small ice chest filled with water and an ice packet. If the travel time is less than one hour, the animal can be left in the plastic bag with no ill effects. For longer trips, the animal can be released into the ice chest. Only one animal may be held in each container; they will fight if they are kept together.

Giant Pacific octopuses are collected opportunistically. The Seattle Aquarium has a large cadre of staff and volunteer divers who report octopus sightings to the collectors, who then return to the site if an animal is needed by the aquarium. The dens are usually evident from their middens, although these octopuses are also frequently found out of their dens. While they are mostly nocturnal and more likely seen at night, divers prefer to catch these large octopuses during the day when they are more visible. Aquarium staff divers use a fish anesthetic, Quinaldine, to persuade the octopuses to leave their dens.

Once divers coerce the octopus out of its den, they wait to try to capture it until it is on sand or gravel or is swimming so it won't have anything solid to grab. Divers wrestle these octopuses into a large mesh diver's bag, usually with some difficulty, as the animals can be large and can cling to rocks with surprising strength. Once brought to the surface, the animal is placed in a large ice chest or an even larger insulated tote for transport. The choice of transport container depends on the size of the animal; an animal up to 25 lb. (11 kg) can be transported in a 40-gallon (150-l) ice chest of chilled seawater. During trips longer than one hour, pure oxygen is bubbled into the chilled water. If transport time is shorter, the water is not oxygenated. Roland reports successfully catching and transporting octopuses up to 80 lb. (36 kg) using these methods. Upon arrival, the animal is transferred to a holding area, where it is weighed and measured. Everyone avoids handling the animals, especially with rough nets, because the octopus's skin is delicate and prone to injury from abrasion.

While working on their doctoral research on octopus lateralization and play in Austria, Ruth Byrne and Michi Kuba drove to Italy approximately twice a year to obtain octopuses. They bought octopuses (mostly common octopuses) from the Naples open-air fish market and the Naples Aquarium. To help defray the trip's cost, they also transported fish for several public aquariums in Austria. After filling their car with containers of fish and octopuses, they drove as fast as possible over the Alps from Naples to Vienna. Luckily, they were never blocked by snow. They used both battery-powered air pumps and air pumps powered by an inverter run off the car's battery. If the weather was warm, they ran the air conditioning full blast to keep the water cold. Near the end of the trip, late in the night, they met the curators of the aquariums in dark parking lots for quick transfer of the fish. The octopuses were acclimating to their new homes in less than fourteen hours, and survival rates were 95 percent.

Transportation time is a critical factor, because time and deteriorating water quality are the enemies. The objective is to provide optimal water quality and stability during the trip while minimizing trip time. If an octopus inks in a small volume of water, such as the amount in a shipping container, the ink will coat the animal's gills and it will suffocate. Ink, like other organic substances, will also cause the water quality to deteriorate. Since inking is an antipredator defense, smelling ink may also increase stress. If you collect your own octopus and it is prone to inking, let it ink as much as possible while in your net in the ocean, where you can easily flush out the ink. If transporting an octopus in containers that you have access to, such as plastic buckets in a car, bring some extra seawater so you can do water changes if the animal inks. A turkey baster can also be used to slurp up viscous ink blobs.

Low temperature acts as an anesthetic and slows down the metabolism of poikilothermic invertebrates such as octopuses. A lower metabolism means the octopus breathes more slowly, uses less oxygen, and excretes less waste. Also, cold water holds more oxygen than warm water, and slows down the breakdown of waste products into ammonia. Although in absolute terms colder water is preferable to warmer water for shipping, avoid any sudden change of temperature or an extreme temperature. Rapid temperature changes cause additional stress. Octopuses shipped via airmail are put in insulated fish boxes so that the water temperature will remain stable when they are subjected to extreme temperature changes, like being left outside on the tarmac during a blizzard or sitting in the back of a dark brown delivery truck in the middle of a summer heat wave. You can include ice packs in the shipping container for species that prefer cold water and for summer shipments. In the field, buckets or coolers containing octopuses can be placed in the shade with a wet towel on top of them. Frequent seawater changes are advisable.

Some scientists use magnesium chloride (MgCl) to anesthetize cephalopods during handling prior to shipping. If you are experienced with this substance, the correct dose of MgCl can reduce inking and lower the stress and metabolism and therefore also the use of oxygen and the production of waste products. But using MgCl can be tricky, since the effective dose seems to vary with species and animal size, and too much is deadly. The NRCC is the only group we are aware of that regularly uses this method.

Animals that eat a lot excrete a lot of waste. In a closed system with no filtration, this waste will rapidly build up and have a negative effect on the water quality. When you are transferring an octopus from one captive environment to another, it is a good idea to not feed the animal twenty-four hours prior to the move. You can also use gentle aquarium-safe chemicals that absorb or neutralize ammonia and stabilize water quality. We have used Stress Coat effectively while transporting octopuses.

Octopuses need well-oxygenated water. In containers such as buckets and coolers placed in the back seat of the car where you can access them during transport, use an air pump to provide oxygen and keep the water circulating. For sealed shipments such as in air transport, octopuses are typically packed in two or three large individual plastic bags filled with one-fifth fresh seawater and four-fifths pure oxygen. These bags are then placed in insulated fish boxes. Most of the shipping cost is based on weight of the water, so using larger bags with extra oxygen provides even more of a buffer for delays and shipping stress at negligible additional cost.

Setting Up a Marine System

Readers interested in setting up a marine aquarium for an octopus should already be familiar with systems. We strongly urge those who have not kept a marine system but wish to keep an octopus, to first start with a simple marine system with fish in order to gain experience. Colorful but hardy fish such as damselfish are good choices for beginners. Those who have gained extensive experience with marine animal husbandry can often just look at a system and tell whether there are problems. There is no shortcut for learning this skill; paying close attention to water quality and to the behavior of the marine animals will help you acquire the knowledge you need.

Water with good quality, which can be made from fresh water and sea salt available at the pet store, is chemically close to the unpolluted seawater you would find in the open ocean. The pH of natural seawater is about 8.2. Natural seawater has over seventy known elements dissolved in it, mostly in trace amounts. Only a few make up 99 percent of all the dissolved salts. They and their abundance by weight are: chloride (Cl) 55.04 percent, sodium (Na) 30.61 percent, sulfate (SO_4) 7.68 percent, magnesium (Mg) 3.69 percent, calcium (Ca) 1.16 percent, and potassium (K) 1.10 percent. There are many trace elements in seawater, such as manganese (Mn), lead (Pb), copper (Cu), strontium (Sr), iron (Fe), iodine (I), and even trace amounts of gold (Au). Most of the elements occur in concentrations expressed in parts per thousand (ppt), parts per million (ppm), or parts per billion (ppb). Although the elements are present in small amounts, most are important for biochemical reactions, the chemistry of life.

In all of the world's oceans, the chemical composition of seawater is virtually the same. Only the total quantity of these elements—the salinity—changes slightly from place to place, and nutrients such as nitrogen compounds are present only in very low levels in natural sea water. These natural fertilizers are produced as waste products and from decomposition, and are dangerous in high levels. The world's oceans are vast and can dilute naturally produced waste products so they are never in high concentrations. Many photosynthetic organisms compete for these chemicals; they are essential fertilizers for primary producers, or plants that make their own food.

In closed systems, the water is recirculated, and up to three different types of filters are used to maintain water quality: mechanical, chemical, and biological filtration. Biological filtration is simply filtration that provides a lot of surface area and good conditions for bacteria to grow. The simplest type of biological filtration is an under-gravel filter—a plastic plate that holds gravel and circulates water through it. The gravel gives bacteria a surface to attach to and the water circulation brings in new water to be filtered. The problem with these inexpensive filters in an octopus tank is that the octopus may find a way to get under the filter and thus out of sight, or may destroy it. Canister filters and wet-dry filters are better choices for biological filtration for an octopus tank, because they also have some mechanical filtration and, if you add carbon, they provide chemical filtration too. If the filters are cleaned regularly, the organics are removed from the system before they break down into ammonia, which reduces the load on the biological filtration. If they are not cleaned, they do little good. Chemical filtration with carbon, resins, and protein skimmers (foam fractionators) also reduces the load on the biological filtration. Instead of mechanically trapping particles that are removed, these directly remove undesired compounds from the water.

For an octopus tank, we advise that some level of all three types of filtration be used in a closed system. We also recommend the use of a protein skimmer, even a small one, which is an excellent way to maintain high oxygen levels, help remove ink, and provide chemical filtration.

Good water quality is nothing magical, but in closed systems, it is an ideal we strive for but never perfectly achieve. Aquariums hold more animal life in a smaller body of water than would exist in the wild. Even with the best filtration, the water in aquariums contains more nutrients from animal waste than natural seawater. When there are too many nutrients in the water, certain pest species are likely to grow and reproduce rapidly. *Aiptasia* (small brown sea anemones), bristle worms, and fast-growing algae are typical indicator organisms of excess nutrients.

In the worst case, a large bacteria bloom can form. In moderation, bacteria help break down the waste products, but when the tank is too polluted with nutrients, the bacteria go into overdrive, clouding the water and using up much of the dissolved oxygen. The resulting drop in oxygen kills more life in the system, which adds more nutrients to the water. The tank may go anoxic (no oxygen) in extreme cases, which will kill all life forms that require oxygen, and animals like an octopus that requires high levels of oxygen will be among the first to die. These conditions are likely to occur in newly set up systems and in systems where too many animals are crammed into a small space.

Different species of animals have different tolerances of deviations from ideal water quality. Damselfish are among the most tolerant to less than optimal water quality. Corals, on the other hand, are much less tolerant of poor water quality. Octopuses are in the middle: they can live in surprisingly high nutrient levels but not as high as damselfish, and they need good oxygen levels and low levels of metals such as copper, which is especially deadly to them and most invertebrates. The necessary water quality in closed systems can be maintained by mechanical, chemical, and biological filtration and regular partial water changes.

Some public aquariums, like Hawaii's Waikiki Aquarium, California's Monterey Bay Aquarium, and Washington's Seattle Aquarium, and some wet laboratories such as the University of Hawaii lab on Coconut Island, the Aquatron at Dalhousie University in Nova Scotia, and the Bermuda Institute of Ocean Sciences, are able to keep some or all of their aquariums in an open system—a system in which natural seawater is pumped in from the ocean, circulated through the system, and then returned to the ocean. Open systems are best when the source of water is clean and stable and you desire the natural seasonal fluctuations in temperature and even plankton. At Monterey Bay Aquarium, they filter the incoming water very little, and the larvae of strawberry anemones and other marine life come in with the water, settle out in the tanks, and make a beautiful natural backdrop.

There are many benefits to open systems, such as little or no need for filtration, high water quality, and no need to wait to establish a bacteria filter. But there are some challenges as well. The water quality is only as good as the source. If the source water is affected by red tide or pollution, that's a serious problem with no easy solution. Most open systems are in areas not prone to red tide, and the owners pipe their water in from fairly deep so it is clean and free of pollution. Conditions will rapidly deteriorate if the supply of incoming water stops for any reason, such as power outages or pump breakdowns. Many large systems have backup generators and pumps in case of such emergencies.

In some rare cases, very cold water full of dissolved air or water compressed in a pump with air can become uncompressed or warmed up to the point that it holds more than the saturation level of dissolved gasses like oxygen and nitrogen. The water will get back to equilibrium—the excess air will start to form small bubbles in the supersaturated water—and if this happens inside the body of an animal, it can be killed. Trickling the water through an off-gassing tower eliminates this problem.

If the water needs to be altered in any way, such as warming it or chilling it to a constant temperature, as is often needed in scientific experiments, open systems are incredibly inefficient. It takes a lot of energy to change the temperature of water; to do this and then flush the water through a system and then down the drain is also very wasteful. These systems can be made a little more efficient by transferring some of the heat or cooling from the wastewater to the incoming water. Still, open systems are best when the local water is reliably clean and has desired seasonal temperatures and other natural fluctuations. The primary advantage of open systems is that water quality can be extremely high and that waste products are literally flushed away.

Most systems are closed ones. In these, the waste products of the marine life build up until they are broken down or otherwise removed from the system. For closed systems, it is important to understand the nitrogen cycle. Many books have been written on maintaining marine aquariums, and most of the better ones spend at least a chapter discussing the nitrogen cycle. In a nutshell, dead animals and plants as well as excrement break down into compounds such as ammonia. Ammonia is the first and most deadly of the nitrogen compounds in the nitrogen cycle. If ammonia builds up, it will kill the animals in the system.

Bacteria in the genera *Nitrosomonas* and *Nitrobacter* serve the marine community well as decomposers. They break down nitrogen compounds by oxidizing them to obtain energy, and they eat toxic waste compounds and excrete less toxic ones. These bacteria don't appear overnight in a new aquarium. Well, actually they do, but in nowhere near the numbers needed. It takes about two weeks for enough *Nitrosomonas* to develop to be able to oxidize ammonia (NH_3) to nitrites (NO_2). Ammonia, being so toxic, prevents *Nitrobacter* bacteria from growing. Only when the *Nitrosomonas* bacteria have reduced the ammonia levels can the *Nitrobacter* bacteria begin to proliferate. They also take two weeks to oxidize the nitrites to nitrates (NO_3). The time it takes to build up enough *Nitrosomonas* and *Nitrobacter* bacteria is referred to as "cycling the system." Nitrates, the end product of the nitrogen cycle, are the least toxic of the three nitrogen compounds. But they will continue to build up in the system in higher and higher concentrations until they are removed. Regular partial water changes are an easy way to keep nitrate levels down. Changing about 25 percent of the aquarium's water every month is standard practice for marine aquariums. Nitrates, like all nitrogen compounds in the nitrogen cycle, are fertilizer for algae and fast-growing animals such as *Aiptasia* that depend on algae symbiosis. Systems high in nitrogen are therefore prone to blooms of algae and other pest species.

A more natural but much more complicated way to remove nitrates is to use an algae turf scrubber. Some advanced hobbyists choose to use the scrubber to harvest and remove algae at regular intervals in order to help remove nitrates and other excess nutrients. No matter how nitrates are removed, all closed systems need time to establish colonies of bacteria. While cycling the system, you should not use chemical filtration (carbon, resins, protein skimmers, or even water changes) to compete with the fledgling bacteria colonies for nitrogen. To start cycling a marine system, you need a source of ammonia. The easiest and modern ethical method is to use chemical sources of ammonia—ammonium hydroxide or ammonium chloride. You will need a marine water testing kit to measure the levels of ammonia, nitrite, and nitrate, which can be purchased at any marine pet store.

You can buy ammonium hydroxide at grocery or hardware stores; get the pure stuff, free of other chemicals such as perfumes and cleaning substances that you don't want to introduce to your aquarium. Check the ingredients list and add some to a small bottle of water and shake; pure ammonia will not form foam. Since the amount of ammonia dissolved in water varies in ammonium hydroxide, most authorities simply suggest adding it until 5 ppm is reached. If you use a new bottle (the ammonia will evaporate if the bottle is left open) of ammonium hydroxide (28 percent ammonia) validated by the American Chemical Society (ACS), you can use the following approximate recipe as a guideline: one drop per gallon of water should get the desired 5 ppm concentration. Alternately, you can buy ammonium chloride (NH_4Cl) in solid (salt) form from scientific laboratory supply stores. Unlike ammonium hydroxide, it has standard purity, and the recipe is more precise: add 0.017 oz. (0.055 g) of ammonium chloride per gallon of water to get the 5 ppm ammonia content.

There are many methods to cycle a tank. One method we like is to maintain a concentration of 5 ppm for the first ten days, 1 ppm for the next ten days, and stop adding ammonia altogether during the third ten-day period. During the end of the first ten-day period, you should see the nitrite levels rise, and during the second and third periods the nitrate levels will rise. A few days before adding the first hardy animals to the system, say on day twenty-eight, do a 50-percent water change, and then attach the chemical filtration system, such as the protein skimmer and carbon or resins if you use them.

There are a number of "magic potions" that claim to help inoculate marine aquariums with bacteria. The products often claim to drastically reduce startup time. Some of these products help but many don't work at all. Even for those that can work, leaving them on the shelf of the aquarium supply store too long may mean that the bacteria are dead. The *Nitrosomonas* and *Nitrobacter* bacteria needed are aerobic, they require oxygen to survive, and they have a limited shelf life if sealed in a bottle. Since there is little quality control in these products and no shelf life printed on the bottle, don't rely on them. You can inoculate the system yourself by seeding it with gravel or the filter wash from a mature system. This approach provides a higher level of starting bacteria in strains that have already proven themselves successful in aquarium conditions. Whatever method you use, there simply is no substitute for patience when setting up closed marine systems.

During the cycling period, only biological filtration should be used. In other words, don't use chemical filters that will remove the ammonia and organics that will break down into ammonia. During this period, you actually want ammonia to develop to high toxic levels, so that you develop a strong population of bacteria to break it down.

After a month, hardy fish can be added to the system. The longer you wait with a biological load such as fish in the aquarium, the more stable it will become. Ideally, it is best to wait three months before removing the fish and adding an octopus. Octopuses eat a lot, grow fast, and produce more waste than a fish of a similar size, so take care not to add more and more biological load to your system until it crashes. Instead, remove the fish, add the octopus, and start using additional filtration while continuing to make regular partial water changes.

In addition to this brief introduction to marine systems, there are many aquarium stores, aquarium societies, books, magazines, and Web sites (such as www.reefs.org or www.reefcentral.com) dedicated to the subject of setting up and maintaining marine aquariums. We encourage readers interested in marine systems to learn more from these sources.

Keeping Your Octopus Healthy

Many people have expressed confusion about the requirements for keeping healthy octopuses. (In the past, there was limited guidance, but over the past decade, more accurate and helpful information has become available. Internet chat groups like TONMO.com and cephgroup are very useful.) Some of the confusion may be related to mysterious deaths that are simply the natural deaths of short-lived animals. Other deaths may be the result of elevated levels of trace elements that are deadly to cephalopods. Many sources state that octopuses demand optimal water quality and that they cannot endure any water pollution. But Roger Hanlon and John Forsythe (1985) have shown that octopuses are surprisingly tolerant of less than optimal nitrogen levels. For example, they found that in five species kept in captivity, no reduction of growth or feeding was noted at a pH as low as 7.5, salinities in the range of 32 to 38 ppt, and both ammonia and nitrite in concentrations of 0.2 ppm on a long-term basis. Similarly, they reported that nitrate concentrations up to 500 ppm did not seem to affect growth or feeding much, if any. However, they mentioned that nitrate concentrations above 100 ppm may affect reproduction. While we are not suggesting that you subject your octopus to these high levels of nitrogen compounds, we do want to make the point that these animals are more hardy than many have believed.

Octopuses and many invertebrates are sensitive to excess trace levels of heavy metals such as copper. Copper can be introduced into marine systems in several ways. It can leach out of copper pipes and into tap water used to make saltwater. Many fish medications are copper based and will quickly kill invertebrates such as octopuses. Octopuses are also sensitive to low dissolved oxygen concentrations. The common octopus will die if the oxygen concentration falls below 0.25 percent (2.5 ml/l). For reference, 100 percent oxygen-saturated seawater would contain approximately 0.5 to 0.7 percent (5 to 7 ml/l) oxygen in a normal aquarium. The decomposition of excessive organics, the oxidation of ammonia to nitrite, and the oxidation of nitrite to nitrate will lower the dissolved oxygen level in a closed system. While excessive amounts of these waste products may not be directly toxic, their breakdown may have a detrimental effect on respiration.

Preventing escape from the aquarium is crucial—the importance of keeping an octopus safely confined can't be overemphasized. For over 2000 years, records have been written of octopuses leaving the water. In 330 BCE, Aristotle wrote that octopuses are the only cephalopods that go on dry land. Pliny the Elder, in his *Naturalis Historia* (circa AD 77), wrote of a giant octopus, supposedly over 600 lb. (300 kg), that stole salted fish from villagers on the shore. Since these first records, there have been countless stories of octopuses getting into and out of all sorts of trouble by leaving the water and escaping their confines.

In more recent times, there have been a number of credible and more than a few incredible stories of captive octopuses escaping their confines. In 1875, Henry Lee wrote extensively about his experiences with octopuses in the first British public aquarium at Brighton. He related the oft-cited story of an octopus that crawled out of the water over to a nearby tank to eat lumpfish before returning to its own tank. Brighton aquarists were left with a puzzling mystery of the disappearing lumpfish, until one morning when the octopus was discovered in the lumpfish tank. This story was popularly received in newspapers of the time and even inspired a poem. Various versions of this story still permeate popular culture as urban myths. In this case, there is a seed of truth to the myths since octopuses do move from their homes to forage for food.

A more recent example of an octopus getting into trouble occurred when a giant Pacific octopus kept at the Cabrillo Marine Aquarium in San Pedro, California, pulled out the plastic pipe serving as a water drain, and died after the water drained out of the tank. Some octopus escape stories are amusing. One of our colleagues was transporting octopuses from Indonesia in an ice chest as carry-on luggage aboard an airplane when one escaped, causing a ruckus among the passengers. The octopus was rescued and survived.

In captivity, some octopus species are more likely to escape than others. We surveyed thirty-eight scientists and public aquarists to find out which species of octopuses were most likely to escape. They reported that common octopuses and giant Pacific octopuses are very prone to escaping. High likeliness to escape was also reported for Caribbean reef octopuses; two-spot and blue-ringed octopuses are less likely to escape. However, we have heard of cases where even these species have escaped captivity. In the case of blue-ringed octopuses, the results could be deadly and not just for the octopus. Keeping captive octopuses contained is critical for their welfare. In the story from the Brighton Aquarium, the octopus returned to its tank every night. However, they often do not make it back to a marine

Escape Artists

In 1993, I was returning from a research cruise to Dry Tortugas National Park in the Florida Keys, when we stopped for the night in Key West. I intended to catch an octopus but didn't have buckets with lids; I did have several buckets and my trusty tropical fish collecting net. A friend came along with me and we headed for the rock jetty that protected the marina from waves. Within 20 minutes, we found an octopus. This one was a good-sized (baseball sized, all balled up, and 2 ft. [0.6 m] across, stretched out) Caribbean reef octopus that had been through a lot. Most of her arms were in various states of regrowth and she had scars. When I first caught her, she quickly found a tiny hole in the dive net and was out. I caught her again and she escaped again. The third time I caught her, I quickly closed the hole with my hand.

She was placed in a bucket with 1 gal. (3 l) of water. Not having a lid, I filled the other bucket two-thirds full of water and slid it into the first bucket to cover it. I carefully inspected this setup, knowing octopuses' strength and ability to escape. The water in the top bucket must have weighted at least 20 lb. (9 kg) (and the crack between the two buckets was no larger than $^{1}/_{12}$ in. (13 mm) at the most, much too small for the beak of an octopus that size to get through. My friend said years later that at the time he thought I was being overly careful in my precautions.

We looked for a smaller octopus with full arms that night but didn't see any more cephalopods. About an hour later, we returned to the buckets. The octopus was gone, escaped even before we got her home. To escape from those buckets, she must have pushed the bucket of water up to enlarge the crack and held it there while she escaped. We would have loved to see how she managed to do that.

I can't help but think that this is exactly what evolution has selected in octopuses, a creature designed to use its flexible body, tricks, and smarts to avoid predation from vertebrates. They are truly masters at getting into and out of tight spaces. Many of us have observed octopuses leave the water in the wild; this is true for giant Pacific octopuses and red octopuses in the northeast Pacific and the Caribbean reef octopus and the common octopus in the Caribbean. Norman reports that *Octopus alpheus* will leave the water to crawl between tide pools. Leaving the water is a normal and natural behavior for some species although it can be lethal in captivity.

—James B. Wood

environment. If out of the water too long, they will become desiccated and die. To escape, an octopus must be able to hold onto a surface and climb out or find a hole large enough to squeeze through. Octopuses' suckers are amazingly strong: a single sucker with a 1/4 in. (6 mm) diameter can hold 5.2 oz. (148 g). An octopus can easily pull many times its body weight and can walk straight up a wall.

There are two main techniques to ensure that captive octopuses remain captive. One is to leave a lip around the top of the tank with a material that an octopus cannot get a grip on. The other is to seal the lid on, making sure there are no spaces the octopus can pry open. The first method is popular with public aquariums. They often use large tanks and display the octopuses in a way that hides the equipment. This approach leaves them room to put a barrier around the top of the tank to keep the octopus in. This barrier must be large enough that the octopus cannot reach across it and grab onto something on the other side, and must be coated with a material that octopus suckers cannot grip. For large octopuses, Astroturf is often used. But this material doesn't work for smaller species with smaller suckers, which are able to grab onto the blades of fake grass. Open-cell foam, screening, and rough synthetic cloth will work. Most home aquariums and research tanks use the seal-it-shut method. Acrylic lids are easy to custom make. Holes can be drilled for the intake and return tubes of pumps and for air lines. Clamps, weights, or Velcro straps can be used to secure the lid to the tank.

Octopuses can escape through almost any pipe, fitting, or filter in the tank that is big enough for them to squeeze through or weak enough for them to pull apart. Octopuses are born with the ability to get into and out of things. Even hatchling octopuses can get into places they don't belong; hatchling octopuses crawled up the inside 1/8-in. (3 mm) air line tubing that was used to send water into the octopus rearing chamber. Octopuses like secure, dark, narrow places for their homes: under-gravel filter beds or behind fiberglass backdrops are ideal. Under-gravel filters should not be used in the octopus tank and all backdrops, pipes, drain holes, and plumbing to and from the filter should be secured, sealed, or covered with material that can't be held onto. Not only does the aquarium and filtration have to be octopus proof, anything else that is added to the system also needs to be secured.

Not only can octopuses hide behind things in tanks, they can also open valves, pull corks out, and open containers, especially if they are large.

Most folks don't have the space required for a giant Pacific octopus, but the curator of fishes and invertebrates at the Oregon Coast Aquarium does. A truly large giant Pacific octopus was on display in a 200,000-gallon (790,000-l) tank; she weighed over 100 lb. (45 kg). The aquarium installed a small ROV (remote operated vehicle) in the tank. For a fee, the public could guide it around the aquarium. According to the manufacturer of the ROV, it takes a long-handled strap wrench to open its 6-in. (15-cm) acrylic dome back port. One morning, the octopus grabbed the ROV, unscrewed

Early Lesson

As a child, I went to the Florida Keys for lobster season every summer with my best friend's family. I never was much good as a fisherman, being too impatient, but could free-dive to catch lobsters, tropical fish, and invertebrates. Back at home, I had prepared a 30-gal. (114-l) tank and was determined to come back with something cool, maybe even an octopus. I caught Houdina, my first octopus, on that trip. She taught me many lessons, most of them the hard way. I got home from the Keys well past midnight on the night before the first day of school. While other students were worried about what clothes to wear, I was up most of the night acclimating the octopus, securing the tank with duct tape and placing four-by-fours on the tank's lid. Fifteen years later, that lid is still warped from the weight.

I was tired and knew I might make a mistake, so I went for overkill on securing the aquarium lid. I could always worry about how to open the tank to feed the octopus and making the tank look nice the next day. When I woke up, the octopus was nowhere to be seen. She had somehow escaped! I frantically tore my room apart looking for her, and was late for school. After school, I raced home and looked all over. Days went by and there was no sign of the now certainly dead octopus anywhere in the house. My dad hadn't said anything . . . yet. Finally, days later, I took my dive light and looked up into the bottom of the tank. There she was, under the under-gravel filter plate—she had escaped from view but not from the tank. That was one of the first lessons I learned: no under-gravel filters in an octopus tank. She was a large octopus, and to this day I have no idea how she got under there without disturbing the gravel.

—James B. Wood

the back port, and proceeded to gut the electronics. After all, she had been given enrichment items like lidded jars, so why not try this? Saltwater flooded into the housing and destroyed the circuit boards and color video camera. Finding no food inside, she released the ROV fairly quickly. Fortunately, the unit was not energized so the animal was not shocked. Only some of the ROV was salvageable.

What to Feed Your Octopus

All octopuses are generalist carnivores. In the wild, they primarily eat crabs and mollusks, but individual octopuses can have preferences. We have had success feeding octopuses live fish, crabs, and shrimp, and we have had success occasionally feeding them thawed, previously frozen crustaceans, fish, and squid. To coax a reluctant octopus to accept frozen shrimp, thaw the shrimp and then skewer them on a shish kabob skewer. Then "swim" them around the tank. The motion usually attracts the attention of the octopus. When a new octopus is adjusting to captivity, we usually feed it live food for a month or so while it settles in, before switching to frozen food.

Newly hatched octopuses will only accept live food. They will not grow on frozen or prepared food. Amphipods or mysid shrimp are ideal first foods for hatchling octopuses. Adults can be often fed a variety of frozen seafood that can be purchased at your local grocery store. Once acclimated, an octopus eats a surprising amount of food. There is some anecdotal evidence that a long-term diet based on freshwater animals, particularly feeder goldfish, can cause liver problems in octopuses. Until further studies are completed, we recommend only offering marine-based foods.

A healthy octopus has a healthy appetite. Newly purchased octopuses may still be shy, but most will quickly learn where the food comes from. An octopus that is offered a live marine shrimp and crab and doesn't eat it for days is likely very sick. How much to feed your octopus? About 5 to 10 percent of its body weight per day is the scientific answer, but then you have to know the weight of your octopus, and obtaining that is a challenge in itself. The best method is to carefully watch how the octopus responds to the food you give it. If the octopus takes but only eats part of what you offer, you are offering too much food and should cut back.

Remember, most species of octopuses are nocturnal, a plus if you are a college student. If you are a night owl, we advise feeding at night; if society forces you to wake up early, try feeding early in the morning. In our experience, feeding at a regular time almost guarantees that you will see your octopus at least once a day, true for most octopus species.

Octopuses should be kept in a tank dedicated to them. They view most fish, crustaceans, and mollusks (including smaller octopuses) as food, though there are a few other animals that can be kept with them. Echinoderms such as sea stars, brittle stars, and sea urchins are usually safe tank mates. Reef tank inhabitants such as corals and sponges should be safe for your octopus, although he or she may not approve of your interior arrangement and will do some redecorating. Large, heavy rocks are less moveable.

Octopuses do not need any specialized spectrum or high-intensity lighting, a definite advantage for those of us who are on a budget or are still confused by the myriad lighting possibilities and their ramifications. Given a choice, an octopus probably prefers no lighting for its tank. But of course we humans do like to view our pets. A compromise of one or two standard inexpensive white fluorescent bulbs will work just fine. We prefer to turn the lights on and off with a timer, to add one more constant to the system.

Enrichment

Enrichment, practiced with large octopuses in the Seattle Aquarium, is the animal husbandry practice of enhancing the quality of captive animal care by identifying and providing species-relevant environmental stimuli necessary for optimal psychological and physiological well-being. Enrichment has been a hot topic in animal care in the past decade. But until recently, enrichment had been only applied to vertebrates, especially mammals. Roland and James have been among the first to argue that all animals live in complex environments, and therefore all animals should be considered for enrichment when they are kept in captivity.

We used the giant Pacific octopus as an example of an invertebrate that would benefit from enrichment. At the time, we didn't have any scientific evidence that enrichment for these invertebrates mattered. However, we argued that in the absence of evidence, it was best to assume that enrichment could benefit all captive animals. Although there has still been no study on the effect of enrichment in octopuses, shortly after we published our arguments for including enrichment for animals other than vertebrates, a 2000 study by Ludovic Dickel proved that cuttlefish raised in enriched environments develop better mental ability than those raised in unenriched environments.

Enrichment comes in many forms. One form is a complex environment with hiding places and things to interact with. The primary predators of octopuses—marine fish, mammals, and birds—are all visual predators. Cephalopod adaptations reflect the pressure of these predators. Providing an environment where the octopus can hide, can dig holes, and move rocks is one way to provide enrichment. Materials safe to use as enrichment and decoration in marine aquariums include live rock (usually corals, from the ocean), limestone, or other aquarium-safe rocks, seashells, PVC pipe caps, and glass jars. Octopuses should never be kept in environments where they cannot hide in a dark lair if they wish.

The size of the aquarium is also important. Small species such as the Florida pygmy octopus can be kept in 10-gallon (38-l) aquariums. Medium-sized octopuses such as the Caribbean reef octopus are best kept in aquariums that are at least 30 gal. (114 l). Larger octopuses should be kept in aquariums that are at least 55 gal. (200 l). The giant Pacific octopus can grow to 100 lb. (45 kg) in captivity. These huge octopuses are too large for most home aquariums, since they should be kept in systems of at least 500 gal. (2000 l). Keep in mind that there are many different species and that size at maturity varies considerably even within a species. Also remember that a young, healthy, well-fed octopus will grow considerably. No harm can be done except to your wallet by having a tank that is too large. During a power outage, a large tank will be depleted of oxygen much slower, and a large volume of water may effectively dilute ink. While smaller tanks than those recommended have been used, especially small sterile tanks used by researchers to successfully raise octopuses, we suspect that the animals' quality of life and development are worse (see plate 38).

Another form of enrichment is offering a variety of food. In the wild, animals have to work for their food—they have to locate, capture, and prepare it. Live food is enriching, since the octopus has to capture and process it. Also, food can be placed in puzzle boxes that the octopus has to pull apart. Recently, many public aquariums have started using both these approaches, to enrich their octopuses and as a display to the public of octopus abilities.

Octopuses are fascinating animals, and with some precautions and patience they can be successfully kept throughout their natural life span in captivity. The large-egged species can even be raised from eggs in captivity.

References

Books, Articles, and Papers

Alexander, R. M. 2003. *Principles of Animal Locomotion*. Princeton, New Jersey: Princeton University Press.

Ambrose, R. F. 1984. Food preferences, food availability and the diet of *Octopus bimaculatus* Verrill. *Journal of Experimental Marine Biology and Ecology* 77: 29–44.

Ambrose, R. F. 1986. Effects of octopus predation on motile invertebrates in a rocky subtidal community. *Pubblicazioni della Stazione Zoologica di Napoli, Marine Ecology Progress Series* 30: 261–272.

Anderson, R. C. 1987. Cephalopods at the Seattle Aquarium. *International Zoo Yearbook* 26: 41–48.

Anderson, R. C. 2004. Mollusk mating behaviors. In *Encyclopedia of Animal Behavior*. Ed. M. Bekoff. Portsmouth, New Hampshire: Greenwood Publishing.

Anderson, R. C., P. D. Hughes, J. A. Mather, and C. W. Steele. 1999. Determination of the diet of *Octopus rubescens* Berry, 1953 (Cephalopoda: Octopodidae), through examination of its beer bottle dens in Puget Sound. *Malacologia* 41: 455–460.

Anderson, R. C., and J. A. Mather. 2002.The packaging problem: Bivalve prey selection and prey entry techniques of the octopus *Enteroctopus dofleini*. *Journal of Comparative Psychology* 121: 300–305.

Anderson, R. C., J. A. Mather, and D. L. Sinn. 2008. Octopus senescence: Forgetting how to eat clams. *Festivus* XL: 55–56.

Anderson, R. C., D. L. Sinn, and J. A. Mather. 2008. Drilling localization on bivalve prey by *Octopus rubescens* Berry, 1953 (Cephalopoda, Octopodidae). *The Veliger* 50 (4): 326–328.

Anderson, R. C., and J. B. Wood. 2001. Enrichment of giant Pacific octopuses: Happy as a clam? *Journal of Applied Animal Welfare Science* 4: 157–168.

Anderson, R. C., J. B. Wood, and R. A. Byrne. 2002. Octopus senescence: The beginning of the end. *Journal of Applied Animal Welfare Science* 5: 275–283.

Anderson, R. C., J. B. Wood, and J. A. Mather. 2008. *Octopus vulgaris* in the Caribbean is a specializing generalist. *Marine Ecology Progress Series* 371: 199–202.

Aristotle, 330 BCE. *History of Animals*. Reprint. Trans. D. M. Balme. Cambridge, Massachusetts: Harvard University Press, 1997.

Basil, J. A., R. T. Hanlon, S. I. Sheikh, and J. Atema. 2000. Three-dimensional odor tracking by *Nautilus pompilius*. *Journal of Experimental Biology* 203: 1409–1414.

Bekoff, M., C. Allen, and G. Burghardt. 2002. *The Cognitive Animal*. Cambridge, Massachusetts: MIT Press.

Boal, J. G. 2006. Social recognition: A top down view of cephalopod behavior. *Vie et Mileu* 56: 69–79.

Boal, J. G., A. W. Dunham, K. T. Williams, and R. T. Hanlon. 2000. Experimental evidence for spatial learning in octopuses (*Octopus bimaculoides*). *Journal of Comparative Psychology* 114: 246–252.

Boletzky, S. v. 1987a. Embryonic phase. In *Cephalopod Life Cycles*, vol. 2. Ed. P. R. Boyle. London, England: Academic Publishers, 5–31.

Boletzky, S. v. 1987b. Juvenile behaviour. In *Cephalopod Life Cycles*, vol. 2. Ed. P. R. Boyle. London, England: Academic Publishers, 45–60.

Boletzky, S. v. 2003. *Idiosepius*: Ecology, biology and biogeography of a mini-maximalist. Report for Workshop #3 of CIAC in Phuket, Thailand.

Borelli, L., and G. Fiorito. 2008. Behavioral analysis of learning and memory in cephalopods. In *Learning Theory and Behavior*, vol. 1. Ed. R. Menzel. New York: Elsevier, 605–627.

Boucher-Rodoni, R., E. Boucher-Rodoni, and K. Mangold. 1987. Feeding and digestion. In *Cephalopod Life Cycles*, vol. II: *Comparative Reviews*. Ed. P. R. Boyle. London, England: Academic Publishers, 85–108.

Boycott, B. B. 1954. Learning in *Octopus vulgaris* and other cephalopods. *Pubblicazioni della Stazione Zoologica di Napoli* 25: 67–93.

Boyle, P. R. 1986. Neural control of cephalopod behavior. In *The Mollusca*. Ed. A. Willows. London, England: Academic Publishers, 1–98.

Boyle, P. R. ed. 1987. *Cephalopod Life Cycles*, vol. II: *Comparative Reviews*. London, England: Academic Publishers.

Boyle, P. R. 1991. *The UFAW Handbook on the Care and Management of Cephalopods in the Laboratory*. New York: Hyperion Books.

Budelmann, B. U. 1994. Cephalopod sense organs, nerves and the brain: Adaptations for high performance and life style. *Marine Behavior and Physiology* 25:13–33.

Burghardt, G. M. 2001. Play: Attributes and neural substrates. In *Handbook of Behavioral Neurobiology*, vol. 13. Ed. E. M. Blass. New York: Klüwer, 317–356.

Byers, J. A. 1997. *The American Pronghorn: Social Adaptations and the Ghosts of Predators Past*. Chicago: University of Chicago Press.

Byrne, R. A., M. Kuba, and U. Griebel. 2002. Lateral asymmetry of eye use in *Octopus vulgaris*. *Animal Behaviour* 64: 461–468.

Byrne, R. A., M. Kuba, C. V. Meisel, U. Griebel, and J. A. Mather. 2006. Does *Octopus vulgaris* have preferred arms? *Journal of Comparative Psychology* 120: 198–204.

Caldwell, R. L. 2000. The blue-ringed octopus. *Freshwater and Marine Aquaria* 23: 8–18.

Camhi, J. M. 1984. *Neuroethology*. Sunderland, Massachusetts: Sinauer.

Caspi, A., and D. J. Bem. 1990. Personality: Continuity and change across the life course. In *Handbook of Personality Theory and Research*. Ed. L. Pervin. New York: Guilford, 549–575.

Chase, R. 2002. *Behavior and Its Neural Control in Gastropod Molluscs*. Oxford, England: Oxford University Press.

Cosgrove, J. A. 1993. In situ observations of nesting female *Octopus dofleini*. *Journal of Cephalopod Biology* 2: 33–46.

Cosgrove, J. A., and N. McDaniel. 2009. *Super Suckers: The Giant Pacific Octopus and Other Cephalopods of the Pacific Coast*. Madiera Park, British Columbia: Harbour Publishing.

Coss, R. G., and R. O. Goldthwaite. 1995. The persistence of old designs for perception. In *Perspectives in Ethology*, vol. 11: *Behavioral Design*. Ed. N. S. Thompson. New York: Plenum Press, 83–148.

Cott, H. 1940. *Adaptive Colouration in Animals*. London, England: Methuen Publishers.

Cousteau, J.-I., and P. Diolé. 1973. *Octopus and Squid: The Soft Intelligence*. Garden City, New York: Doubleday.

Cuvier, G. 1829. Memoire sur un ver parasite d'un nouveau genre (*Hectocotylus octopodis*). *Annals de Science Naturelle* 18: 247–158.

Davis, H., and D. Balfour. 1992. *The Inevitable Bond: Examining Scientist-Animal Relationships*. Cambridge, England: Cambridge University Press.

Dews, P. B. 1959. Some observations on an operant in the octopus. *Journal of Experimental Analysis of Behavior* 2: 57–63.

Dickel, L., J. G. Boal, Bernd U. Budelmann. 2000. The effect of early experience on learning and memory in cuttlefish. *Developmental Psychobiology* 36 (2): 101–110.

Dickel, L., A. S. Darmaillacq, R. Poirier, V. Agin, C. Bellanger, and R. Chichery. 2006. Behavioral and neural maturation in the cuttlefish *Sepia officinalis*. *Vie et Milieu* 56 (2): 89–95.

Dodge, R., and D. Scheel. 1999. Remains of the prey: Recognizing the midden piles of *Octopus dofleini* (Wülker). *The Veliger* 42: 260–266.

Ellis, R. 1998. *The Search for the Giant Squid.* Guilford, Connecticut: Lyons Press.

Fiorito, G., and F. Gherardi. 1999. Prey-handling behaviour of *Octopus vulgaris* (Mollusca, Cephalopoda) on bivalve prey. *Behavioural Processes* 46: 75–88.

Fiorito, G., C. von Planta, and P. Scotto. 1990. Problem solving ability of *Octopus vulgaris* Lamark (Mollusca, Cephalopoda). *Behavioral and Neural Biology* 53: 217–230.

Flash, T., and B. Hochner. 2005. Motor primitives in vertebrates and invertebrates. *Current Opinion in Neurobiology* 15: 660–666.

Forsyth, A. 2001. *A Natural History of Sex.* Buffalo, New York: Firefly Books.

Forsythe, J. W., and R. T. Hanlon. 1997. Foraging and associated behavior by *Octopus cyanea* Gray, 1849, on a coral atoll, French Polynesia. *Journal of Experimental Marine Biology and Ecology* 209: 15–31.

Gallup, G. G. Jr., J. R. Anderson, and D. J. Shillito. 2002. The mirror test. In *The Cognitive Animal.* Eds. M. Bekoff, C. Allen, and G. M. Burghardt. Cambridge, Massachusetts: MIT Press, 325–333.

Gosling, S. D. 2001. From mice to men: What can we learn about personality from animal research? *Psychological Bulletin* 127: 45–86.

Grasso, F. W. 2008. Octopus sucker-arm coordination in grasping and manipulation. *American Malacological Bulletin* 24: 13–24.

Griffin, D. R. 1981. *The Question of Animal Awareness.* New York: Rockefeller University Press.

Gross, M. G. 1976. *Oceanography.* Columbus, Ohio: Merrill.

Hall, K. C., and R. T. Hanlon. 2002. Dynamics of the mating system of the giant Australian cuttlefish *Sepia apama* Gray. *Bulletin of Marine Science* 71: 1125.

Hanlon, R. 2007. Cephalopod dynamic camouflage. *Current Biology* 17 (11): 400–404.

Hanlon, R. T., L.-A. Conroy, and J. W. Forsythe. 2008. Mimicry and foraging behavior of two tropical sand-flat octopus species off North Sulawesi, Indonesia. *Biological Journal of the Linnean Society* 93: 23–38.

Hanlon, R. T., and J. W. Forsythe. 1985. Advances in the laboratory culture of octopuses for biomedical research. *Laboratory Animal Science* 35: 33–40.

Hanlon, R. T., J. W. Forsythe, and S. v. Bolelzky. 1985. Field and laboratory behavior of "macrotritopus larvae" reared to *Octopus defilippi* Veranyi, 1851 (Mollusca: Cephalopoda). *Vie et Mileu* 35: 237–242.

Hanlon, R. T., J. W. Forsythe, and D. E. Joneschild. 1999. Crypsis, conspicuousness, mimicry and polyphenism as antipredator defences of foraging octopuses on Indo-Pacific coral reefs, with a method of quantifying crypsis from video tapes. *Biological Journal of the Linnean Society* 66: 1–22.

Hanlon, R. T., and J. B. Messenger. 1988. Adaptive coloration in young cuttlefish (*Sepia officinalis* L.): The morphology and development of body patterns and their relation to behavior. *Philosophical Transactions of the Royal Society of London* B 320: 437–487.

Hanlon, R. T., and J. B. Messenger. 1996. *Cephalopod Behaviour.* Cambridge, England: Cambridge University Press.

Hazlett, B. A. 1995. Behavioral plasticity in Crustacea: Why not more? *Journal of Experimental Marine Biology and Ecology* 193: 57–66.

Hirschfeld, L. A., and S. A. Gelman. 1994. Toward a topography of mind: An introduction to domain specificity. In *Mapping the Mind: Domain Specificity in Cognition and Culture.* Eds. L. A. Hirschfeld and S. A. Gelman. Cambridge, England: Cambridge University Press, 3–35.

Hochberg, F. G. 1990. Diseases of Mollusca: Cephalopoda. In *Diseases of Marine Animals,* vol. III. Ed. O. Kinne. Hamburg, Germany: Biologische Anstalt Helgoland, 47–228.

Hochberg, F. G. 1998. Class cephalopoda. In *Taxonomic Atlas of the Benthic Fauna of the Santa Maria Basin and the Western Santa Barbara Channel,* vol. 8. Eds. P. V. Scott and J. A. Blake. Santa Barbara, California: Santa Barbara Museum of Natural History, 175–236.

Hockett, C. F., and S. A. Altmann. 1968. A note on design features. In *Animal Communication.* Ed. T. A. Sebeok. Bloomington: Indiana University Press, 61–72.

Huffard, C. L., R. L. Caldwell, and F. Boneker. 2008. Mating behavior of *Abdopus aculeatus* (d'Orbigny, 1834) (Cephalopoda: Octopodidae) in the wild. *Marine Biology* 154: 353–367.

Hvoreckny, L. M., et al. 2007. Octopuses (*Octopus bimaculoides*) and cuttlefishes (*Sepia pharaonis, S. officinalis*) can conditionally discriminate. *Animal Cognition* 10: 449–459.

Iglesias-Garcia, J., et al. 2007. Rearing of octopus paralarvae: Present status, bottlenecks and trends. *Aquaculture* 266: 1–15.

Kagan, J. 1994. *Galen's Prophecy: Temperament in Human Nature.* New York: Basic Books.

Kier, W. M., and A. M. Smith. 1990. The morphology and mechanics of octopus suckers. *Biological Bulletin* 178: 126–136.

Kier, W. M., and K. K. Smith. 1985. Tongues, tentacles and trunks: The biomechanics of movement in muscular-hydrostats. *Zoological Journal of the Linnean Society* 83: 307–324.

Kölliker, A. v. 1845. Some observations upon the structure of two new species of Hectotyle parasitic upon *Tremoctopus violaceus* Della Chiaje and *Argonauta argo* L., with an exposition of the hypothesis that these Hectocotylae are males of the Cephalopoda upon which they are found. *Transactions of the Linnaean Society of London* 20: 9–21.

Kováks, Á., and J. A. Mather. 2008. Cephalopod cognition and scholastic psychology. *Res Cogitans* 5: 23–38.
Kuba, M., R. A. Byrne, D. V. Meisel, and J. A. Mather. 2006. When do octopuses play? The effect of repeated testing, age and food deprivation on object play in *Octopus vulgaris*. *Journal of Comparative Psychology* 120: 184–190.
Kubodera, T., and K. Mori. 2005. First-ever observation of a live giant squid in the wild. *Proceedings of the Royal Society B: Biological Sciences* 272 (1581): 2583–2586.
Lane, F. W. 1957. *Kingdom of the Octopus*. London, England: Jarrolds Publishers.
Lange, M. M. 1921. On the regeneration and finer structure of the arms of the cephalopods. *Journal of Experimental Zoology* 31 (1): 1–55.
Langridge, K. V., M. Broom, and D. Osorio. 2007. Selective signalling by cuttlefish to predators. *Current Biology* 17: 1044–1045.
Lee, H. 1875. *The Octopus*. London, England: Chapman & Hall.
Leite, T. S., M. Harmovici, J. A. Mather, and J. F. Lino Oliviera. 2009. Habitat, distribution, and abundance of the commercial octopus (*Octopus insularis*) in a tropical oceanic island, Brazil: Information for management of an artisanal fishery inside a marine protected area. *Fisheries Research* 98: 85–91.
Leite, T. S., and J. A. Mather. 2008. A new approach to octopuses' body pattern analysis: A framework for taxonomy and behavioral studies. *American Malacological Bulletin* 24: 31–42.
Lorenz, K. Z. 1961. *King Solomon's Ring*. Trans. M. K. Wilson. London, England: Methuen.
Lutz, R. A., and J. R. Voight. 1994. Close encounter in the deep. *Nature* 371: 563.
Mangold, K., and F. G. Hochberg. 1991. Defining the genus *Octopus*: Redescription of *Octopus vulgaris*. *Bulletin of Marine Science* 49: 665.
Mann, T. 1984. *Spermatophores*. Berlin: Springer-Verlag.
Marshall, N. J., and J. B. Messenger. 1996. Colour-blind camouflage. *Nature* 382: 408–409.
Mason, W. 1984. Learning: Physiology and behavior. *Verhandlungen der Deutsche Zoologische Gesellschaft* 77: 45–56.
Mather, J. A. 1988. Daytime activity of juvenile *Octopus vulgaris* in Bermuda. *Malacologia* 29: 69–76.
Mather, J. A. 1991a. Foraging, feeding and prey remains in middens of juvenile *Octopus vulgaris* (Mollusca: Cephalopoda). *Journal of Zoology* (London), 224: 27–39.
Mather, J. A. 1991b. Navigation by spatial memory and use of visual landmarks in octopuses. *Journal of Comparative Physiology* A 168: 491–497.
Mather, J. A. 1992. Interactions of juvenile *Octopus vulgaris* with scavenging and territorial fishes. *Marine Behaviour and Physiology* 19: 175–182.
Mather, J. A. 1994. "Home" choice and modification by juvenile *Octopus vulgaris*

(Mollusca: Cephalopoda): Specialized intelligence and tool use? *Journal of Zoology* (London), 233: 359–368.

Mather, J. A. 1995. Cognition in cephalopods. *Advances in the Study of Behavior* 24: 316–353.

Mather, J. A. 1998. How do octopuses use their arms? *Journal of Comparative Psychology* 112: 306–316.

Mather, J. A. 2001. Animal suffering: An invertebrate perspective. *Journal of Applied Animal Welfare Science* 4: 151–156.

Mather, J. A. 2008a. Cephalopod consciousness: Behavioral evidence. *Consciousness and Cognition* 17: 37–48.

Mather, J. A. 2008b. To boldly go where no mollusc has gone before: Personality, play, thinking, and consciousness in Cephalopods. *American Malacological Bulletin* 24: 51–58.

Mather, J. A. In press. Octopus. In *Encyclopedia of Animal Behaviour.* Eds. M. D. Breed and J. Moore. Oxford, England: Academic Press.

Mather, J. A., and R. C. Anderson. 1993. Personalities of octopuses (*Octopus rubescens*). *Journal of Comparative Psychology* 107: 336–340.

Mather, J. A., and R. C. Anderson. 1999. Exploration, play, and habituation in *Octopus dofleini. Journal of Comparative Psychology* 113: 333–338.

Mather, J. A., and R. C. Anderson. 2007. Ethics and invertebrates: A cephalopod perspective. *Diseases of Aquatic Organisms* 75: 119–129.

Mather, J. A., and R. C. Anderson. In press. Octopuses (*Enteroctopus doflieni*) learn to open jars with cross-modal cueing. *Ferrantia.*

Mather, J. A., and D. M. Logue. In press. The bold and the spineless: Approaches to personality in invertebrates. In *Animal Personalities: Behavior, Physiology, and Evolution.* Eds. C. Carera and D. Maestripieri. Chicago, Illinois: University of Chicago Press.

Mather, J. A., and D. L. Mather. 2004. Apparent movement in a visual display: The Passing Cloud in *Octopus cyanea. Journal of Zoology* (London) 263: 89–94.

Mather, J. A., D. L. Mather, and M. Wong Chang. 1997. Cross-species associations of *Octopus cyanea* Gray 1843 (Mollusca: Cephalopod). *The Veliger* 40: 174–177.

Mather, J. A., and R. K. O'Dor. 1991. Foraging strategies and predation risk shape the natural history of juvenile *Octopus vulgaris. Bulletin of Marine Science* 49: 256–269.

Maurer, C., and D. Maurer. 1988. *The World of the Newborn.* New York: Basic Books.

Menzel, R. 1985. Learning in honeybees in an ecological and behavioral context. *Experimental Behavioral Ecology* 31, 55–74.

Messenger, J. B. 1977. Prey capture and learning in the cuttlefish, *Sepia.* In *The Biology of Cephalopods.* Eds. M. Nixon and J. B. Messenger. London, England: Academic Publishers, 347–376.

Messenger, J. B. 2001. Cephalopod chromatophores: Neurobiology and natural history. *Biological Review* 76: 473–528.

Moynihan, M. H., and A. F. Rodaniche. 1982. The behaviour and natural history of the Caribbean reef squid *Sepioteuthis sepioidea*, with a consideration of social, signal and defensive patterns for difficult and dangerous environments. *Advances in Ethology* 125: 1–150.

Müller, H. 1852. Note upon the male of the argonaut and the hectocotylus. *Annals and Magazine of Natural History* 54: 492–493.

Muntz, W.R.A. 1999. Visual systems, behaviour, and environments in cephalopods. In *Adaptive Mechanisms in the Ecology of Vision*. Eds. S. N. Archer et al. Dordrecht, Netherlands: Klüwer, 467–483.

Neisser, U. 1976. *Cognitive Psychology*. New York: Appleton-Century-Crofts.

Nixon, M., and P. N. Dilly. 1977. Sucker surfaces and prey capture. In *The Biology of Cephalopods*. Eds. M. Nixon and J. B. Messenger. London, England: Academic Publishers, 447–511.

Norman, M. 2000. *Cephalopods: A World Guide*. Hackenheim, Germany: Conch Books.

O'Dor, R. K. 1998. Life history strategies. In *Squid Recruitment Dynamics: Influences on Variability within the Genus Illex*. Eds. P. Roidhouse, E. G. Dawe, and R. K. O'Dor. Food and Agricultural Organization Fisheries Technical Paper 376: 233–254.

O'Dor, R. K., and D. M. Webber. 1986. The constraints on cephalopods: Why squid aren't fish. *Canadian Journal of Zoology* 64: 1591–1605.

Olton, D. S. 1979. Mazes, maps and memory. *American Psychologist* 34: 583–596.

Packard, A. 1972. Cephalopods and fish: The limits of convergence. *Biological Review* 47: 241–307.

Packard, A. 1988. The skin of cephalopods (coleoids): General and special adaptations. In *The Mollusca: Form and Function*, vol. 11. Eds. K. M. Wilbur and M. R. Clarke. New York: Academic Publishers, 37–67.

Packard, A. 1995. Organization of cephalopod chromatophore systems: A neuromuscular image-generator. In *Cephalopod Neurobiology*. Eds. N. J. Abbott, R. Williamson, and L. Maddock. Oxford, England: Oxford University Press, 331–367.

Papini, M., and M. Bitterman. 1991. Appetitive conditioning in *Octopus cyanea*. *Journal of Comparative Psychology* 105: 107–114.

Parker G. A. 1984. Evolutionary stable strategies. *Behavioural Ecology: An Evolutionary Approach*. 2nd ed. Eds. J. R. Krebs and N. B. Davies. Oxford, England: Blackwell, 30–61.

Randall, J. E. 1967. Food habits of reef fishes of the West Indies. *Studies in Tropical Oceanography* 5: 665–847.

Renner, M. J. 1990. Neglected aspects of exploratory and investigatory behavior. *Psychobiology* 18, 16–22.

Roper, C.F.E., M. J. Sweeney, and C. Nauen. 1984. *Cephalopods of the World.* FAO Fisheries Synopsis. No. 125, vol. 3. Rome: Food and Agriculture Organization of the United Nations.

Ross, D. M., and S. v. Boletzky. 1979. The association between the pagurid *Dardanus arrosor* and the actinian *Calliactus parasitica:* Recovery of activity in "inactive" *D. arrosor* in the presence of cephalopods. *Marine Behavior and Physiology* 6: 175–184.

Rowell, C.H.F. 1963. Excitatory and inhibitory pathways in the arm of *Octopus. Journal of Experimental Biology* 40: 257–260.

Schaller, G. B. 1972. *The Serengeti Lion: A Study of Predator-Prey Relations.* Chicago: University of Chicago Press.

Scheel, D. 2002. Characteristics of habitats used by *Enteroctopus dofleini* in Prince William Sound and Cook Inlet, Alaska. Pubblicazioni della Stazione Zoologica di Napoli I. *Marine Ecology* 23: 185–206.

Scheel, D. A., A. Lauster, and T.L.S. Vincent. 2007. Habitat ecology of *Enteroctopus dofleini* from middens and live prey surveys in Prince William Sound, Alaska. In *Cephalopods Present and Past: New Insights and Fresh Perspectives.* Ed. N. H. Landsman. New York, Springer, 434–458.

Scheumack, D. D., M.E.H. Howden, I. Spence, and R. J. Quinn. 1978. Maculotoxin: A neurotoxin for the venom glands of the octopus *Haplochlaena maculosa* identified as tetrodotoxin. *Science* 199: 188–189.

Seino, S., Y. Tsuchiya, M. Tsychiya, K. Katayama, and T. Hamada. 1994. Spawning behavior in *Octopus vulgaris* revealed by endoscope observation: How to form egg masses. Paper presented at the symposium on the Behavior and Natural History of Cephalopods, Vico Equense, Italy.

Sinn, D. L., and N. A. Moltschaniwskyj. 2005. Personality traits in dumpling squid (*Euprymna tasmanica*): Context-specific traits and their correlation with biological characteristics. *Journal of Comparative Psychology* 119: 99–110.

Sinn, D. L., N. A. Perrin, J. A. Mather, and R. C. Anderson, R. C. 2001. Early temperamental traits in an octopus (*Octopus bimaculoides*). *Journal of Comparative Psychology* 115: 351–364.

Solem, A. 1974.*The Shell Makers: Introducing Mollusks.* New York: Wiley.

Steenstrup, J. 1857. Several particulars about the giant buttle-fishes of the Atlantic Ocean. *The Cephalopod Papers of Japetus Steenstrup, A Translation into English.* New edition. Eds. A. Volsoe, J. Knudsen, and W. Rees. Copenhagen, Denmark: Danish Science Press, 1962.

Steer, M. A., and J. M. Semmens. 2003. Pulling or drilling, does size or species matter? An experimental study of prey handling in *Octopus dierythraeus* (Norman 1992). *Journal of Marine Biology and Ecology* 290: 165–178.

Steinbeck, J. 1951. *The Log from the Sea of Cortez.* New York: Viking.
Stephens, D. W., and J. R. Krebs. 1986. *Foraging Theory.* Princeton, New Jersey: Princeton University Press.
Strickland, R. M. 1983. *The Fertile Fjord.* Seattle: Washington Sea Grant.
Sutherland, N. S. 1960. Theories of shape discrimination in *Octopus. Nature* 186: 840–844.
ten Cate, W. J. 1928. Contribution a l'innervation des ventouses chez *Octopus vulgaris. Archives Neerlandaises de Physiologie de l'Homme et des Animaux* 13: 407–422.
Thomas, A., and S. Chess. 1977. *Temperament and Development.* New York: Brunner/Mazel.
Thompson, J. T., and J. R. Voight. 2003. Erectile tissue in an invertebrate animal: The octopus copulatory organ. *Journal of Zoology* 261: 101–108.
Tinbergen. N. 1972. *The Animal in its World.* Cambridge, Massachusetts: Harvard University Press.
Van Heukelem, W. F. 1983. *Octopus cyanea.* In *Cephalopod Life Cycles,* vol. I: *Species Accounts.* Ed. P. R. Boyle. London, England: Academic Press.
Vermeij, G. J. 1993. *A Natural History of Shells.* Princeton, New Jersey: Princeton University Press.
Villanueva, R. 1994. Decapod crab zooea as food for rearing cepholopod paralarvae. *Aquaculture* 128: 143–152.
Villanueva, R., and M. D. Norman. 2008. Biology of the planktonic stages of benthic octopuses. *Oceanography and Marine Biology Annual Review* 46: 105–202.
Vincent, T.L.S., D. Scheel, and K. R. Hough. 1997. Some aspects of diet and foraging behavior of *Octopus dofleini* (Wülker, 1910) in its northernmost range. Pubblicationi della Stazione Zoologica di Napoli: *Marine Ecology* 19, 13–29.
Voight, J. R., and A. J. Grehan. 2000. Egg brooding in the deep-sea octopuses in the North Pacific Ocean. *Biological Bulletin* 198: 94–100.
Voight, J. R., H. O. Pörtner, and R. K. O'Dor. 1994. A review of ammonia-mediated buoyancy in squids (Cephalopoda: Teuthoidea). In *Physiology of Cephalopod Mollusca.* Eds. H. O. Pörtner, R. K. O'Dor, and D. L. Macmillan. Amsterdam, Netherlands: Gordon & Breach, 193–204.
Von Frisch, K. 1967. *Bees, Their Vision, Chemical Senses and Language.* Ithaca, New York: Cornell University Press.
Ward, P. D. 1987. Natural History of *Nautilus.* London, England: Allen & Unwin.
Warnke, K., R. Soller, D. Blohm, and U. Saint-Paul. 2004. A new look at geographic and phylogenetic relationships within the species group surrounding *Octopus vulgaris* (Mollusca: Cephalopoda): Indications of very wide distribution from mitochondrial DNA sequences. *Journal of Zoological Systematics and Evolutionary Research* 42: 306–312.

Wells, M. J. 1978. *Octopus: Physiology and Behavior of an Advanced Invertebrate.* London, England: Chapman & Hall.

Wells, M. J., and J. Wells. 1970. Observations on the feeding, growth rate, and habits of newly settled *Octopus cyanea. Journal of Zoology* (London) 161: 65–74.

Williamson, R., and A. Chrachri. 2004. Cephalopod neural networks. *Neurosignals* 13: 87–98.

Wodinsky, J. 1969. Penetration of the shell and feeding on gastropods by *Octopus. American Zoologist* 9: 997–1010.

Wodinsky, J. 1973. Ventilation rate and copulation in *Octopus vulgaris. Marine Biology* 20: 154–64

Wodinsky, J. 1977. Hormonal inhibition of feeding and death in *Octopus*: Control by optic gland secretion. *Science* 198: 948–51.

Wood, J. B., and R. C. Anderson. 2004. Interspecific evaluation of octopus escape behavior. *Journal of Applied Animal Welfare Science* 7: 95–106.

Wood, J. B., and R. C. Anderson. 2009. Keeping captive cephalopods. *Marine Fish and Reef USA* 11: 18–31.

Wood J. B., E. Kenchington, and R. K. O'Dor. 1998. Reproduction and embryonic development time of *Bathypolypus arcticus*, a deep-sea octopod (Cephalopoda: Octopoda). *Malacologia* 39 (1–2): 11–19.

Wood, J. B., and R. K. O'Dor. 2000. Do larger cephalopods live longer? Effects of temperature and phylogeny on interspecific comparisons of age and size at maturity. *Marine Biology* 136 (1): 91–99.

Wood, J. B., K. E. Pennoyer, and C. D. Derby. 2008. Ink is a conspecific alarm cue in the Caribbean reef squid, *Sepioteuthis sepioidea. Journal of Experimental Marine Biology and Ecology* 367: 11–16.

Wood, J. B., and D. A. Wood. 1999. Enrichment for an advanced invertebrate. *The Shape of Enrichment.* 8 (3): 1–5.

Young, J. Z. 1971. *The Anatomy of the Nervous System of* Octopus vulgaris. Oxford, England: Clarendon.

Web Sites

www.thecephalopodpage.org/: *The Cephalopod Page* is J. B. Wood's personal Web site dedicated to cephalopods. Among the longest running animal sites, it has been featured in *Science* three times since its creation in 1995.

www.cephbase.utmb.edu/: *CephBase*, 1998–2004. Created by J. B. Wood, C. Day, and R. K. O'Dor as part of the Census of Marine Life, the site was one of the world's first online species-level databases. The site is no longer in the hands of its creators, and is not currently functional.

http://www.mapress.com/mr/content/v25/2005f/n2p070/: This site describes the mimic octopus (*Thamoctopus mimicus*).

http://tech.groups.yahoo.com/group/ceph/: *Cephgroup* is a listserv associated with *The Cephalopod Page*. This e-group is dedicated to cephalopods and is a chat group open to everyone.

www.reefcentral.com: *Reef Central* is an online community where quality information about the marine and reef aquarium hobby can be exchanged among all levels of hobbyist, from beginner to advanced. The site's goals are to help educate people about the saltwater aquarium hobby and to enhance awareness of the fragility of coral reefs around the world.

www.reefs.org: Dedicated to aquarium hobbyists, this site has a mission to foster responsible marine aquarium hobbyists through education.

www.tonmo.com: *The Octopus News Magazine Online* was created by Tony Morelli in 2000. The site is committed to being a resource for all things cephalopod, and has excellent international community involvement.

Index